Eugene Benjamin Wilson

A Treatise on Practical and Theoretical Mine Ventilation

Eugene Benjamin Wilson

A Treatise on Practical and Theoretical Mine Ventilation

ISBN/EAN: 9783743390140

Manufactured in Europe, USA, Canada, Australia, Japa

Cover: Foto ©Lupo / pixelio.de

Manufactured and distributed by brebook publishing software (www.brebook.com)

Eugene Benjamin Wilson

A Treatise on Practical and Theoretical Mine Ventilation

A TREATISE

ON

PRACTICAL AND THEORETICAL

MINE VENTILATION.

BY

EUGENE B. WILSON,

INSTRUCTOR IN DRIFTON INDUSTRIAL SCHOOL FOR MINERS
AND MECHANICS.

NEW YORK:
JOHN WILEY & SONS,
15 Astor Place.
1887.

PREFACE.

THE author, aware of the interest manifested by miners in this subject, which is so intimately connected with their daily employment, has endeavored to deal with ventilation in such a manner that no one with a fair knowledge of the English language and of arithmetic need despair of thoroughly mastering it.

Knowing that the miner possesses but a comparatively small stock of words, and is not an adept in algebraic formulas, the writer has taken a different position from the standard works on the subject, endeavoring to do away with abstruse language, and such highly mathematical formulas as are only calculated for well-educated engineers. In order that the text may be more readily followed, each article is illustrated by an example. From the number of fatal explosions which have taken place within the past eighteen months, it is evident that either managers are ignorant of the laws of ventilation, or else negligent in providing methods in conformity with those laws. If the latter is the case, they are the more blamable; as ignorance of the subject, even at this time, may be excusable, but negligence can never be. There are many practical hints given for engineers, who, owing to lack of time, have been unable to keep as well informed on this subject as they may have wished to do: also many useful memoranda, and tables for saving time and labor when dealing with questions relating to ventilation, will be found in this little volume.

The author would also state, that this book is intended for miners, and not for engineers; but at the same time it is believed that it will be found useful to many engineers.

In hopes of showing how men's lives may be lengthened, and preserved from the dangers that lurk in improperly ventilated mines, this work has been written, and is now offered to the class most interested, by their friend and well-wisher,

E. B. W.

DRIFTON, PENN., June, 1884.

CONTENTS.

CHAPTER I.

		PAGE.
§ 1.	Atmosphere. — Specific Gravity of Gases	1
§ 2.	Nitrogen	5
§ 3.	Oxygen	7
§ 4.	Carbonic Acid. — Black-Damp	9
§ 5.	Carbonic Oxide. — White-Damp	12
§ 6.	Sulphuretted Hydrogen	13
§ 7.	Marsh-Gas. — Fire-Damp	14
§ 8.	Expansion of Gases	19
§ 9.	Falling Bodies	22

CHAPTER II.

§ 10.	Natural Ventilation	23
§ 11.	Motive-Column. — Head of Air	25
§ 12.	Variation of Temperature, Effects of	27

CHAPTER III.

§ 13.	Safety-Lamps	29
§ 14.	Comparison of Safety-Lamps and their Efficiency	37
§ 14a.	Detection of Fire-Damp	39

in conformity with those laws. If the latter is the case, they are the more blamable; as ignorance of the subject, even at this time, may be excusable, but negligence can never be. There are many practical hints given for engineers, who, owing to lack of time, have been unable to keep as well informed on this subject as they may have wished to do: also many useful memoranda, and tables for saving time and labor when dealing with questions relating to ventilation, will be found in this little volume.

The author would also state, that this book is intended for miners, and not for engineers; but at the same time it is believed that it will be found useful to many engineers.

In hopes of showing how men's lives may be lengthened, and preserved from the dangers that lurk in improperly ventilated mines, this work has been written, and is now offered to the class most interested, by their friend and well-wisher,

E. B. W.

DRIFTON, PENN., June, 1884.

CONTENTS.

CHAPTER I.

		PAGE.
§ 1.	Atmosphere. — Specific Gravity of Gases	1
§ 2.	Nitrogen	5
§ 3.	Oxygen	7
§ 4.	Carbonic Acid. — Black-Damp	9
§ 5.	Carbonic Oxide. — White-Damp	12
§ 6.	Sulphuretted Hydrogen	13
§ 7.	Marsh-Gas. — Fire-Damp	14
§ 8.	Expansion of Gases	19
§ 9.	Falling Bodies	22

CHAPTER II.

§ 10.	Natural Ventilation	23
§ 11.	Motive-Column. — Head of Air	25
§ 12.	Variation of Temperature, Effects of	27

CHAPTER III.

§ 13.	Safety-Lamps	29
§ 14.	Comparison of Safety-Lamps and their Efficiency	37
§ 14a.	Detection of Fire-Damp	39

vi CONTENTS.

CHAPTER IV.
PAGE.
§ 15. Physical Properties of Air in Motion 40
§ 16. Perimeter, Area, etc., defined 42
§ 17. Friction of Air in Airways 43
§ 18. Water-Gauge 43
§ 19. Co-efficient of Friction 47
§ 20. Pressure 49
§ 21. Pressure, continued 50

CHAPTER V.
§ 22. Laws of Pressure, and Friction of Air in Mines, with Problems illustrating the Same 53

CHAPTER VI.
§ 23. Laws regulating the Quantity of Air flowing through Mines 59
§ 24. Problems illustrating above Laws 61
§ 25. Water Gauge and Pressure 67

CHAPTER VII.
§ 26. Ventilation of Single Pits or Drifts 69
§ 27. Ventilation of Mines with Two Orifices . . . 72
§ 28. Doors and Regulators 73

CHAPTER VIII.
§ 29. Splitting the Air 76
§ 30. Explanation of Splits 77

CONTENTS. vii

		PAGE.
§ 31.	Advantages of Splits	79
§ 32.	Area of Airways	85
§ 33.	Problem on splitting the Air	86

CHAPTER IX.

§ 34.	Quantity of Air necessary for a Mine	88
§ 35.	Prevention of Explosions	92
§ 36.	Efficient Ventilation	96

CHAPTER X.

§ 37.	Air Measurements	97
§ 38.	Method of Procedure	101
§ 39.	Barometer	103

CHAPTER XI.

§ 40.	History of Mechanical Ventilators	109
§ 41.	Fans	112
§ 42.	Guibal Fan	115
§ 43.	Comparative Economy between Furnace and Fan Ventilation	117
§ 44.	Formulas	121
§ 45.	Problems	124
§ 46.	Quantity of Air necessary per Man	128
§ 47.	Treatment of Asphyxiated Persons	130

MINE VENTILATION.

CHAPTER I.

1. THE atmosphere is composed of nitrogen and oxygen, with a trace of carbonic-acid gas.

These three gases, essential to the existence of all animal and vegetable life, when taken separately will not support life. A mechanical mixture of these gases, in the proportion of four parts of nitrogen to one part of oxygen, is the air we breathe; which, if mixed with deleterious gases (or, as we say, impure), will cause serious physical disorders, and not unfrequently premature death. Carbonic-acid gas rarely exceeds one part in sixteen hundred of pure air; being present in the atmosphere, so say our best chemists, in the ratio of four parts in ten thousand. The term "atmosphere" designates that immense expanse or ocean of gaseous matter which envelops or surrounds our earth, commonly called "air." It is supposed that this atmosphere is forty-five miles thick about the earth: which, however, is merely supposition, as the height has not as yet been computed with accuracy, although it has been proven that Mariotte's law is conformed to by the gases which constitute the air; their density varying according to the pressure.

That the atmosphere varies in pressure was recognized at an early period: even the Florentine pump-makers were acquainted with the fact that water could not be raised by suction from a depth of more than thirty to thirty-three feet.

Galileo explained this phenomenon, and clearly demonstrated that the pressure of the atmosphere was equal to the weight of thirty-three feet of water.

Torricelli argued from this, that, if the atmosphere would support thirty-three feet of water, it would not support more than thirty inches of mercury; as mercury is about fourteen times heavier than water. The result of Torricelli's investigations and experiments gave us the instrument known as the barometer, by means of which we can measure the density of the atmosphere, which is on an average equal to the weight of a column of mercury thirty inches in height at sea-level. The temperature of the atmosphere is not the same throughout: it becomes colder as we ascend; hence on the top of high mountains we find snow the year round.

That air has weight may be shown by the following experiment. Take a vessel whose capacity, say, is 100 cubic inches, exhaust it of air, and then weigh it. Let it now be filled with dry air at the ordinary temperature and pressure, then weighed again. Upon second weighing it is found to be 31,074 grains heavier than at the first. As the weight of the atmosphere will sustain a column of mercury whose base is one inch square, and whose height is thirty inches, it must press down with a weight equal to the weight of the mercury of the above dimensions to balance it. The weight of this mercury is 14.7225 pounds, and hence the atmosphere has a weight or pressure equal to 14.7 on each square inch of surface exposed to it.

Air is taken as the standard of comparison for all gases and vapors. The chemical composition of air in its natural state is given by Dr. Frankland as follows:—

Oxygen 20.61
Nitrogen 77.95
Carbonic acid 04
Moisture 1.40
                              ———
                              100.00

Dry air is composed of

	Per cent by weight.	Per cent by vol.
Nitrogen	77	79
Oxygen	23	21
	100	100

The specific gravity of the gases in the following table was determined experimentally by De la Roche and Berad, who took air as the standard for gases.

TABLE I.

Name of Gas.	Symbols.	Specific gravity.
Atmosphere.	N_4O	1.0000
Hydrogen	H	0.0692
Water-vapor	H_2O	0.6210
Olefiant gas	C_2H_4	0.9672
Carbonic oxide	CO	0.9674
Nitrogen	N	0.9713
Nitric oxide	NO	1.0390
Oxygen	O	1.1056
Nitrous oxide	N_2O	1.5250
Carbonic acid	CO_2	1.5290
Sulphurous acid	SO_2	2.470

Manner of finding the weight of a gas compared with that of air is illustrated by the following problem.
Example. If 1,000 cubic feet of air weigh 80.728 pounds when the temperature is 32° and the barometer 30″, what will 1,000 cubic feet of nitrogen weigh under the same conditions?

Solution.— From the table we find the density of nitrogen to be 0.9713 when air is one: hence we have the proportion 1 : 0.9713 = 80.728 : *Ans.*, or 78.415 + pounds.

§ 2. MINE VENTILATION. 5

2. The miner has to deal with several gases, especially the coal-miner. It is therefore imperative that he should know the composition of these gases, so as to be able to distinguish them. To this end, therefore, a brief synopsis of some of the gases so often met with in mines, together with their properties, is here inserted.

Nitrogen.
Symbol, N. Equivalent, 14. Specific gravity, 0.9713.

The name signifies nitre-maker. It constitutes about four-fifths of the atmosphere, and enters into a great variety of combinations. Nitrogen is somewhat lighter than air; a cubic foot of the gas weighing 0.0784167 pounds, while a cubic foot of air weighs 0.080728 pounds. Nitrogen may be obtained by burning the oxygen from a confined portion of air. It is incapable of sustaining combustion or animal life: not that it has positive poisonous properties; but flame is extinguished, and animals smother, for want of oxygen. It is best characterized by its passiveness; as it has very little affinity or attraction for other elements, and upon the slightest provocation will free itself if possible. For instance, it may be induced to combine with iodine, and form "nitric iodide," a black, insoluble powder, which will explode if moved, jarred, or even touched with a feather. It enters into the composition of gunpowder, nitro-glycerine, and dynamite. With oxygen it forms five distinct compounds: —

    a, Nitrous oxide, N_2O.
    b, Nitric oxide, NO.
    c, Nitrous anhydride, N_2O_3.
    d, Nitrogen peroxide, NO_2.
    e, Nitric anhydride, N_2O_5.

a, Nitrous oxide, or nitrogen sub-oxide, when pure, may be respired for a few minutes with impunity. When inhaled in large quantities, it produces a lively intoxication, accompanied with violent laughter: whence it derives the name of "laughing-gas."

b, Nitric oxide, or nitrogen protoxide, may be prepared by treating copper filings, or turnings, with nitric acid. The gas obtained in this manner is colorless and transparent: in contact with air or oxygen, it produces deep-red fumes.

c, Nitrous anhydride is an obscure body, and forms, with the elements of water, an acid known as "nitrous acid."

d, Nitrogen peroxide is the chief constituent of the deep-red fumes noticed when nitrogen protoxide is brought in contact with air.

e. Nitric anhydride is a very unstable, white, solid compound, decomposing spontaneously into nitrogen peroxide and oxygen. When treated with water, it forms nitric acid.

Oxygen.

Symbol, O.   Equivalent, 16,   Specific gravity, 1.1056.

3. Oxygen forms one-fifth part of the atmosphere. It is transparent and colorless, not to be distinguished by its aspect or smell from atmospheric air. It is the most widely diffused of all the elements, forming about one-third of the solid crust of the globe. It unites with all the other elements to form compounds, which are sometimes gaseous, sometimes solid, sometimes liquid. The name signifies acid-former; and, with one exception, oxygen enters into the combination of acids. All the ordinary phenomena of fire and light which we daily witness depend upon the union of the body burned with the oxygen of the air: in fact, the term "oxidation" may, for all ordinary purposes, be regarded as synonymous with "combustion."

Faraday has roughly estimated that the amount of oxygen required daily to supply the lungs of the human race is at least one thousand millions of pounds; that required for the respiration of the lower animals is at least twice as much as this; while the always active process of decay requires certainly no less than four times as much. Faraday also estimates that one thousand millions of pounds are sufficient to sustain all the artificial fires lighted by man, from the camp-fire of the savage to the roaring blaze of the blast-furnace, or the raging flames of a grand conflagration.

Amount of Oxygen required Daily.

	Pounds.
Whole population	1,000,000,000
Animals	2,000,000,000
Combustion and fermentation	1,000,000,000
Decay and other processes	4,000,000,000
Total amount of oxygen required daily	8,000,000,000

These figures are inconceivable; and, when we reduce the oxygen consumed to tons, we fail to grasp it, as it is no less than 3,571,428 tons.

Although the consumption of oxygen is so great, yet there is no fear of its being exhausted; as, at the present rate of consumption, there is enough to last nine hundred thousand years. Oxygen is the active principle of the atmosphere. It devours every thing with which it can unite: it corrodes metals, decays fruits, promotes combustion, and is a prime necessity for health.

The body is a stove, in which fuel is burned; the chemical action being the same as in any other stove. We take into our lungs air, and give out a poisonous gas,—carbonic-acid, the waste products of the combustion of our bodies. From this we may learn how important a factor oxygen is for health, and how necessary it is that we have plenty of fresh, pure air, if we wish to be free from disease. One man breathes into his lungs at each inspiration about 230 cubic inches of air, or one gallon. In the delicate cells of the lungs the air gives up its oxygen to the blood, receiving, in turn, carbonic acid and water, foul with waste matter which the blood has picked up in its circulation through the body. Should we rebreathe it into our lungs, the blood will leave the lungs, not bearing invigorating oxygen, but refuse matter to obstruct the whole system. Without oxygen the muscles become inactive, the heart acts slowly, food is undigested, brain is clogged, and at last such fatal results as were manifested in the "Black Hole of Calcutta" implore us not to be stingy or afraid of "God's blessing,"—pure air.

4. Having examined slightly the constituent parts of the atmosphere, let us briefly examine the principal gases met with in coal-mining.

Carbonic-Acid Gas.

Symbol, CO_2. Equivalent, 22. Specific gravity, 1.53.

One cubic foot of the gas at 32° F., and barometer at 30″, weighs 0.12845 of a pound.

This gas is composed of carbon and oxygen. Miners have given it several names, such as "stythe," "choke-damp," "black-damp," and "after-damp." This gas is always produced when compounds containing carbon are burnt in air or oxygen. It may be produced by treating limestone or marble with hydrochloric or sulphuric acids. The occluded gases in all coal contain carbonic acid. Carbonic acid is considered poisonous, on account of the many deaths which have resulted from burning charcoal and carbonaceous materials in places where there was a deficiency of ventilation, and by reason of the fatal nature of after-damp of explosions in coal-mines. Its specific gravity is 1.524; so that it is a little more than one and a half times as heavy as air. It lodges near the floor of places in which it is evolved when little more than mutual diffusion is going on. Owing to its great density, it may be poured from one vessel to another. It is the only gas, except nitrogen, which is evolved by most bituminous coals; and, when it is given off in quantity, active ventilation is required to carry it off.

Le Blanc, and many other chemists, affirm that air containing more than five parts in a thousand is injurious to breathe. Mr. J. W. Thomas of England, while not asserting that it is not injurious, says, that "in levels and seams of semi-bituminous and bituminous coals in South Wales, in part or wholly worked to the dip, with scanty ventilation in some particular spots, through the non-completion of air-splits, or conveyances, men often work in an atmosphere containing from two to five per cent of this gas for hours." Be that as it may, the system, uninspired by the energizing oxygen, is sensitive to cold. The pale cheek, the lustreless eye, the languid step, shortness of breath, speak but too plainly of oxygen starvation. "In such a soil, catarrh, scrofula, miners' asthma, and consumption run riot."

Miners call the carbonic acid produced by the explosion of fire-damp, "after-damp." They fear it almost as much as fire-damp, as it instantly destroys the lives of all who may have escaped the flames of the explosion. This property of carbonic acid, of choking or smothering, has of late years been made use of for putting out fires in coal-mines. In one case, an English mine which had been burning twenty years was smothered by pouring into it eight billion cubic feet of carbonic acid, and then closing it up for one month. At the end of the month the mine was opened, and found to be ready for the resumption of labor.

When found alone in a mine, carbonic acid is not considered as dangerous as fire-damp, since it will not burn. Carbonic acid, at the ordinary temperature and pressure, is a gas. It solidifies when subjected to great pressure; but, as soon as the pressure is removed, it returns to the gaseous state: therefore the term "carbonic acid" is applied as well to the gas as to the acid.

Carbonic Oxide.

Symbol, CO.   Equivalent, 14.   Specific gravity, 0.9674.

5. One cubic foot of the gas at 32° F., and barometer of 30″, weighs 0.078305 of a pound.

This gas is sometimes called "white-damp." From experiments made by Dr. Meyer and J. W. Thomas, it was found, that, during every explosion, large quantities of this gas were formed, and that the fatal effects of the after-damp are in a great measure due to its presence. Carbonic oxide is an odorless and colorless gas, incapable of supporting the combustion of other bodies, but is itself an inflammable gas. It possesses very poisonous properties, which act powerfully on the blood and nervous system, producing, when inhaled in very small quantities, a most unpleasant sensation, followed quickly by headache, and disinclination to move, prostration and inactivity: if continued to be breathed, asphyxia follows, and death soon results. Air containing only one-half per cent of this gas would prove fatal, if inspired for any length of time. Mr. Thomas advocates oxygen and induced artificial breathing, for those who are overcome by this gas, in preference to the administration of alcoholic stimulants.

The composition of carbonic oxide is,

	By weight.	By volume.
Carbon	42.86	1
Oxygen	57.14	1
	100.00	2

This gas is narcotic, and, when breathed in a concentrated form, would produce no pain, the body passing instantly into a state of coma. Whatever position the victim assumed, in that position he would be found dead, unless moved by some other means. "Carbonic acid, and the nitrogen left after an explosion, would be fatal in their effects; but very often men have succumbed to supposed after-damp, while their lamps burned well. The presence of carbonic acid and nitrogen will not account for the result or phenomena."

SULPHURETTED HYDROGEN.

Symbol, SH_2. Equivalent, 17. Specific gravity, 1.178.

6. One cubic foot of the gas at 32° F. and barometer of 30″, weighs 0.09492 of a pound.

This gas, although not common, is met with sometimes in mines. It is colorless, but easily distinguished by its peculiar smell,—that of rotten eggs. It may be prepared by treating sulphide of iron with dilute sulphuric acid. The composition of sulphuretted hydrogen is,

	By weight.	By atoms.
Sulphur	94.12	1
Hydrogen	5.88	2
	100.00	3

When mixed with oxygen, it will explode if ignited. When inhaled in a pure state, it is a powerful narcotic poison, and produces fainting and asphyxia when present in very small proportions of the atmosphere. It appears to be probable that the gas is generated in small quantities in old worked-out mines. Some claim that it is formed by the decomposition of pyrites in old workings; others, that it is not formed in this manner, but by the decomposition of props and timber standing in water, by breaking up the sulphate of lime, and assimilating its oxygen, while sulphur seizes upon the hydrogen of the wood to form sulphuretted hydrogen. This gas is also known as hydrosulphuric and sulphuric acid gas.

MARSH-GAS.

Symbol, CH_4. Equivalent, 8. Specific gravity, 0.55314.

7. One cubic foot of this gas, at 32° F. and barometer of 30″, weighs 0.044665 of a pound.

It is known by several names, — proto-carburetted hydrogen, light carburetted hydrogen, hydride of methyl, fire-damp. Marsh-gas, however, is better known to

§ 7.   MINE VENTILATION.   15

miners as "fire-damp." It is colorless, tasteless, odorless, when pure, burning with a yellowish flame. It is formed in swamps and marshy places by the decomposition of vegetable matter, and may be seen bubbling up through the water when the mud is stirred beneath. Marsh-gas is found in such quantities in some places, that it is used for lighting towns, and evaporating brine. In the oil-regions it frequently bursts forth with explosive violence, throwing the oil high in the air when the drill nears it. Coal-gas contains about thirty-eight per cent of marsh-gas. When marsh-gas is evolved in the shape of "blowers," it constitutes about ninety-six per cent of the total volume. Blowers sometimes assume enormous dimensions, and have been conveyed from the workings to the surface by means of pipes, and utilized. Marsh-gas is not poisonous. Sir H. Davy, of safety-lamp fame, was the first to experiment on this gas. He found, that, when mixed with three and a half times its volume of air, it did not explode; with five and a half times its volume, it exploded slightly; and, when mixed with eight or nine volumes of air, the force of explosion was greatest. When there is a deficiency of ventilation, the fire-damp is said to rise to the upper portion or top of a gallery, and there remain, because of its being lighter than air. It is also said that carbonic acid, being heavier than air, lodges on the "floor" or "thill" of a mine. Mr. Thomas, in his book on "Mines, Gases, and Ventilation," says, "This impression is erroneous; and while not denying that fire-damp is often found in larger quantities near the roof, and carbonic acid in larger quantities near the floor, these positions do not prove that they have lodged there, nor is it so: on the contrary, marsh-gas is always diffusing in every direction; and it is only in those places where the gas is evolved in greater quantity than will diffuse, or become carried away by the ventilation, that accumulation takes place. These erroneous ideas in reference to marsh-gas arose from the fact that it is found in the crevices and holes, and near the roof, of coal-mines." The explanation of this is very simple. The fact of portions of top-rock falling, and the squeezing-in or lowering of the top throughout the whole worked portion above the coal, affords communication with, it may be, some rider or unworked seam of coal above, with the receding working-face, or with crevices which are in communication with stores of fire-damp extending to considerable distances. Now, the pressure of the atmosphere being the same on all sides, the gas in the fissures and cracks in the top are subjected to that pressure externally, so that air finds no outlet through these cracks, and the diffusion which takes place is simply "natural" or mutual diffusion.

The fire-damp issuing into these cracks, etc., encounters the same pressure as if issuing direct into the air-current; so that it would find its way downwards by virtue of the extra force of pressure of the imprisoned gas into the top of the holes, goaves, or receptacles in the top-rock, and it would be more likely to escape or find an exit in these places, or come in contact with the ventilating current here, owing to the fact that more easy communication is afforded by the dislocation and partial disintegration of the top caused by a fall, followed by a partial opening-up of the surrounding mass. The fire-damp, therefore, instead of accumulating and lodging at the roof of the gallery by virtue of its lesser density, is forced downwards until it finds its way into the holes and other receptacles, and is continually fed from above.

"When marsh-gas escapes from the floor, and makes its appearance in quantity near the roof, it shows deficiency of ventilation, or a strong outpour of gas; but even then, if there is any ventilation at all, it will be largely mixed with air."

The reason that fire-damp is in the form of an explosive mixture in holes in the top, is because it is fed more quickly from above than it is able to diffuse; the ventilating current not affecting it to a very great extent.

Let the lines below represent a gallery from which some of the top has fallen. F is the floor, RR the roof, and H the hole caused by the fall. If a lamp be raised about halfway between R and F, an explosive mixture may be encountered; but, if the lamp is held near the

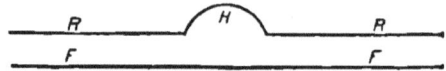

cracks in RR, not a trace of gas will be indicated by it, provided there is sufficient ventilation. Gradually stop the ventilating current, so that the air scarcely travels, and then apply the lamp, and the fire-damp issuing from the cracks may be detected. Active ventilation sweeps away all the gas which escapes from cracks in the line of the top rock or roof; but as the current can find no outlet through the hole at H, and encounters a pressure equal or superior to that along the roof, it travels onward, heedless of any gas situated out of the line RR. When fire-damp accumulates near the roof or top line of a gallery or heading, it indicates that the gas is given off in greater quantity than can be carried off by the ventilation; and, where any such accumulation takes place, there must be either a deficiency of ventilation, or an unusual inpour of gas.

EXPANSION OF GASES.

8. One of the chief characteristics of any gas is its expansive property. To calculate the expansion of any volume of air, the starting-point must always be taken at 0° on Fahrenheit's scale; for air at that temperature will expand $\frac{1}{459}$ of its volume for every degree of heat added. Therefore 459 cubic feet of air at 0° will become 460 cubic feet at 1° F.

Careful experiments show that 459 cubic feet of air at 0° F. weigh 39.76 pounds, when the pressure is 30 inches of mercury of the density due to 32°, — a pressure equal to 14.7 pounds per square inch; but, when the pressure is one inch, it weighs only $\frac{1}{30}$ part of this, or 1.3253 pounds. To find the weight of a cubic foot of air at any temperature or height of the barometer, let

$B =$ height of the barometer in inches,
$t =$ temperature by Fahrenheit's thermometer;

then

(1) $$W = \frac{1.3253 \times B}{459 + t}.$$

Problem. — What is the weight of a cubic foot of air, the temperature of which is 96°, under a barometric pressure of 29.5 inches of mercury?

By substituting 29.5 in formula 1 for B, and 96° for

t, and then performing the operations indicated, we have

$$W = \frac{1.3253 \times 29.5}{459 + 96} = 0.07044 \text{ pound}$$

as the weight of a cubic foot of air under the above conditions.

The following table has been made out to facilitate calculations. It gives the weight of 100 cubic feet of air in pounds at different barometrical pressures.

§ 8. MINE VENTILATION. 21

TABLE II.

Temperature, Fahrenheit.	Weight of 100 cubic feet of air in lbs. due barometer 29".	Weight of 100 cubic feet of air in lbs., due to bar. 29.5".	Weight of 100 cubic feet of air in lbs., due to bar. 30".	Weight of 100 cubic feet of air in lbs., due to bar. 30.5".	Weight of 100 cubic feet of air in lbs., due to bar. 31".	Difference in weight of 100 cu. ft. of air, due to rise and fall of barometer, .5".
30	7.857	7.993	8.129	8.265	8.400	0.136
32	7.826	7.961	8.096	8.231	8.366	0.135
42	7.670	7.802	7.934	8.066	8.198	0.132
52	7.519	7.649	7.779	7.909	8.039	0.130
62	7.375	7.502	7.629	7.756	7.883	0.127
72	7.236	7.361	7.486	7.611	7.736	0.125
82	7.103	7.225	7.347	7.471	7.593	0.122
92	6.974	7.094	7.214	7.334	7.454	0.120
102	6.849	6.968	7.085	7.204	7.323	0.119
112	6.729	6.845	6.961	7.077	7.193	0.116
122	6.614	6.728	6.842	6.956	7.070	0.114
132	6.502	6.614	6.726	6.838	6.950	0.112
142	6.394	6.504	6.614	6.728	6.834	0.110
152	6.289	6.397	6.506	6.614	6.722	0.108
162	6.188	6.294	6.401	6.508	6.614	0.106
172	6.089	6.195	6.300	6.405	6.509	0.104
182	5.995	6.098	6.201	6.304	6.408	0.103
192	5.903	6.004	6.106	6.208	6.309	0.101
202	5.813	5.913	6.014	6.114	6.214	0.100
212	5.726	5.825	5.924	6.023	6.122	0.099
222	5.642	5.739	5.837	5.934	6.032	0.097
232	5.561	5.657	5.753	5.843	5.939	0.096
242	5.481	5.576	5.671	5.765	5.861	0.095
252	5.404	5.497	5.591	5.684	5.777	0.093
262	5.329	5.421	5.513	5.605	5.697	0.092
272	5.256	5.347	5.438	5.527	5.618	0.091
282	5.185	5.275	5.364	5.453	5.542	0.089
292	5.117	5.205	5.293	5.381	5.469	0.088
302	5.049	5.136	5.223	5.310	5.397	0.087

9. As we have seen, air has weight, and therefore becomes subject to the "physical laws" that govern liquids and falling bodies; i.e., air is acted on by gravity in the same manner as a solid.

Let $h =$ the distance fallen through in feet.
$v =$ the velocity acquired at the end of the fall, in feet per second.
$g =$ the distance in feet which an unresisted gravitating body falls in the first second of time; which distance has been found by experiment to be $16\tfrac{1}{12}$ feet near the earth's surface.

Since a body falls $16.08'$ in one second, it gains a velocity of $32.16'$ at the end of the first second: hence we have

(2) $$v = \sqrt{2gh} = 8.0208\sqrt{h}.$$

In this equation h represents the necessary height, in feet, of a vertical air-column which will produce by its weight a velocity equal to v. If this velocity be represented in feet per minute, we shall have

(3) $$v = 8.0208\sqrt{h} \times 60 = 481.2\sqrt{h},$$

and

(4) $$h = \frac{v^2}{(481.2)^2} = \frac{v^2}{231{,}600}.$$

These formulæ (2), (3), and (4) are only theoretically true as regards air in mines, the pressure or head of air-column required being from ten to twenty times as much in order to overcome the friction and resistance of airways, etc., in underground workings; and, were it not for these resistances, very small pressure would suffice to produce great velocity.

CHAPTER II.

NATURAL VENTILATION.

10. MOTION in air is caused by pressure, or difference of pressure. When air becomes heated, it ascends, because it assumes a larger volume; and, as the same volume of cold air is heavier, it pushes the warmer up, or out of its place. This phenomenon in the open air gives rise to winds and breezes, which vary in intensity, according as the cool air takes the place of the warm air, rapidly or slowly.

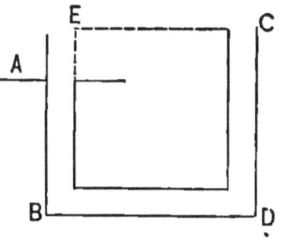

Suppose we have two shafts (connected, as in the above figure) of equal depths, the air in them having the same temperature and density throughout. If no artificial means be used to cause a movement in either direction, — down one, and up the other, — the air will be in equilibrium, and be stagnant. If the tops of the shafts be not of the same height, as AB, the air AE, if of the same density and temperature, would have the same effect as if the shaft rose to E, and remain in equilibrium. Suppose, now, that the external air is of lower temperature than the air in the shafts, and the column AB lower in temperature than CD, a movement of the air then takes place from A towards C. On the contrary, if the external air be warmer than that of the shafts, the shaft AB will be the upcast, and circulation will be from C to A, on account of the heavier column of air CD.

This is the principle upon which furnaces are employed to ventilate mines. They are placed at the bottom, or near the bottom, of the upcast shaft. The heat given off by them raises the temperature of the air, which expands $\frac{1}{459}$ part of its volume for each degree of heat added, and hence becomes lighter. The cooler air of the downcast shaft is now able, owing to its greater density, to fall down the shaft, and push the air of the mine into the upcast, which becomes, in turn, heated by the furnace. Let us take an example to see in what manner this difference in density will cause

§ 11. MINE VENTILATION. 25

ventilation. Suppose the two shafts to be of equal depth, say 900 feet, and suppose the barometer to stand at 29 inches; also let the temperature in the downcast be assumed to be 42°, and that of the upcast 72°. From the table on p. 21, we find that 100 cubic feet of air at 42° weigh 7.67 pounds: therefore 900 cubic feet weigh 9 times as much, or 69.030 pounds. Again: we find 100 cubic feet of air at 72° weigh 7.236 pounds, and hence 900 cubic feet weigh 65.124 pounds. The difference between these weights will give the pressure that would cause circulation.

Pounds.
900 cubic feet at 42° = 69.030
900 cubic feet at 72° = 65.124
——————
Difference in weight 3.906

This pressure in the open air would produce a velocity of wind between twenty-five and thirty miles per hour.

11. The "motive-column" is a "head of air" of such a height that it will equal the difference between the weight of the downcast and upcast columns of air.

Let M = the motive-column, or head of air,
D = the depth of the upcast in feet,
t = the temperature of the upcast in degrees,
t_1 = the temperature of the downcast in degrees,

then,

(5) $$M = D \times \frac{t-t_1}{459t + t_1}$$

Problem.— Find the motive-column which would produce a pressure of 3.906 pounds when the upcast has a depth of 900 feet and a temperature of 72°, and the downcast has the same depth and a temperature of 42°.

Substituting in formula 5 we have

$$M = 900 \times \frac{72 - 42}{459 + 42} = 53.892'$$

as the length of the motive-column.

The height of a column of this air, one foot in area, weighing one pound, may be found by dividing the motive-column by the pressure per square foot; thus:—

$$\frac{53.8922}{3.906} = 13.8'.$$

The relative diameters of the shafts make no difference upon the total pressure, so far as the considerations regarding ventilation are concerned. This is termed "the pneumatic paradox;" for if we have one square foot area for a downcast, and an upcast of ten square feet, the pressure on each square foot of upcast will be the same as on the one square foot of the downcast, provided there is no friction.

§ 12. MINE VENTILATION. 27

When calculating the weight of the motive-column, we must bear in mind that it is not the total amount of air, or the total number of cubic feet, that we seek, but the length of an air-column which has for its base one square foot, and which weighs a certain amount for each cubic foot in height.

12. The temperature of the atmosphere varies at different seasons of the year. This variation causes considerable increase and decrease in the ventilation of mines. To illustrate this, suppose we have a mine 900 feet deep, what will be the ventilating pressure, when the temperature in the downcast is, on an average, 42°, and in the upcast 202°, the barometer being 30″ on an average in the two shafts, which are equal in length?

Pounds.

900 cubic feet of air at 42° F. = 71.406
900 cubic feet of air at 202° F. = 54.126

Difference in weight . . . 17.280

Suppose, now, the temperature of the downcast to be raised to 82° F., then

Pounds.

900 cubic feet of air at 82° F. = 66.123
900 cubic feet of air at 202° F. = 54.126

Difference in weight. . . . 11.997

$17.280 - 11.997 = 5.283$ pounds of the total pressure lost by this change of temperature. As the temperature does not change so much as 40°, in a short time it will not have so marked an effect upon the ventilation of mines: still, an increase of power will be required during the summer months in order to keep the ventilation uniform during the year.

From the above, it may be seen that natural ventilation may be more during the winter than during the summer months. Where furnaces or steam-jets are employed to produce ventilation, the longer the upcast, the better; as the longer upright column of light air gives rise to a brisker ventilation. For this reason, furnaces should not be used in shallow pits.

When coal is worked at a dip, the effect of natural ventilation is very complicated; and in many instances there will be little benefit derived therefrom at any season of the year, owing to the tendency of the heated air to ascend against the down current. Natural ventilation is not a help to artificial ventilation, but often is of very great hinderance to a fan, on account of the vacillating atmospheric changes. On this account, therefore, the inlet and outlet should be as nearly as practicable on the same level. High winds, directed by hills, blowing against the exhaust duct of a fan, greatly impede its action.

TABLE III.
VELOCITY AND POWER OF WINDS (SMEATON).

Velocity. Miles per hour.	Perpendicular force on one square foot, in pounds avoirdupois.	Common appellation of such winds.
1	.005	Hardly perceptible.
4	.079	Gentle wind.
5	.123	
10	.492	Pleasant breeze.
15	1.107	
20	1.968	Very brisk.
25	3.075	
30	4.429	High wind.
35	6.027	
40	7.873	Very high.
50	12.300	Storm.
60	17.715	Great storm.
80	31.490	Hurricane.
100	49.200	Violent hurricane.

CHAPTER III.

SAFETY-LAMPS.

13. IN the year 1814 Mr. Buddle, an Englishman, read a paper before a society formed for preventing accidents in coal-mines, illustrating the various modes employed in the ventilation of collieries by plans and sections.

At that time the only light used in coal-mines was

the candle made of sheep or ox tallow, the latter being considered the better.

When the air in the mine became mixed with inflammable gas, the mode of determining its existence and degree of inflammability was described by Mr. Buddle as follows: —

"In the first place the candle, called by the colliers 'the low,' is trimmed; that is, the liquid fat is wiped off, the wick snuffed short, and carefully cleansed of red cinders, so that the flame may burn as purely as possible. The candle, being thus prepared, is holden between the fingers and thumb of the one hand; and the palm of the other hand is placed between the eye of the observer and the flame, so that nothing but the spire of the flame can be seen, as it gradually towers over the upper margin of the hand. The observation is generally commenced near the floor of the mine, and the light and hand are gently raised upwards till the true state of the circulating current is ascertained. The first indication of the presence of inflammable air is a slight tinge of blue, a bluish-gray color, shooting up from the top of the spire of the candle, and terminating in a fine extended point. This spire increases in size, and receives a deeper tinge of blue, as it rises, through an increased proportion of inflammable gas, till it reaches the firing-point. The experienced collier knows all the gradations of shew (as it is called), and seldom fires the inflammable gas, except in cases of sudden discharge."

When the air was highly charged with inflammable gas, the steel mill was resorted to. It consisted of a steel wheel, to which was applied a piece of flint when it was turned rapidly, thus throwing off a continuous succession of sparks, the light of which was rather uncertain: however, it was a substitute. But as it required one man to work the mill for every man cutting coal, mining became too expensive; and only those portions of the mine were worked where a sufficient current of air could be brought to bear upon the gas to dilute it sufficiently to allow of candles being used.

In 1814 Dr. Clanny produced a lamp by which a light could be used in an inflammable mixture of gas with impunity. The insulation of the flame was accomplished by means of water; and, although the first lamp which was produced, it was too complicated and cumbrous for general use.

In 1815 — at the same time, but in distant localities — Mr. George Stephenson and Sir Humphry Davy both produced lamps which insulated lights in inflammable mixtures of fire-damp without exploding the gas externally. These productions have been of the utmost importance in coal-mining, and consequently to the commercial interests of the country generally.

Mr. Stephenson reasoned, that "if a lamp could be made to contain the burnt air above the flame, and to permit the fire-damp to come in below in a small quantity, to be burned as it came in, the burnt air would prevent the passing of the explosion upwards; and the velocity of the current of the air from below would also prevent it passing downwards." He accordingly constructed a lamp of tin, with a hole in the bottom to admit the air to the flame, and a top perforated with holes. By experiments with this lamp he discovered the true principles of the safety-lamp.

Sir H. Davy, at about the same time, communicated with a friend that he had "discovered that explosive mixtures of mine-damp will not pass through small apertures or tubes, and that if a lamp or lantern be made air-tight on the sides, and furnished with apertures to admit the air, it will not communicate flame to the outward atmosphere." He subsequently found that "iron-wire gauze, composed of wires from one-fortieth to one-sixtieth of an inch in diameter, and containing twenty-eight wires, or seven hundred and eighty-four apertures to the inch, was safe under all circumstances."

The process by which Mr. Davy arrived at the above conclusion is given by himself in a small work "On the Safety-Lamp for Coal-Mines, with some Researches on Flame:" "In reasoning upon the various phenomena brought about by my various experiments, it occurred to me, — as considerable heat was required for the inflammation of the fire-damp, and as it produced, in burning, a comparatively small degree of heat, — that the effect of carbonic acid or azote, and of the surfaces of the small tubes in preventing its explosion, depends upon their cooling powers, or their lowering the temperature of the exploding-mixture so much that it was no longer sufficient for its continuous inflammation."

Mr. Stephenson's lamp has been much improved. It consists of a glass cylinder above the lamp, covered by a cylinder of wire gauze; and, instead of air passing through the perforated plate, it passes through the meshes of the gauze (Fig. 1).

The Davy lamp differs from the Stephenson, inasmuch as the former admits air through the meshes of the wire on all sides: consequently, when immersed in an inflammable mixture, the whole cylinder becomes filled with flame, and ultimately the wires become red-hot. Yet they radiate sufficient heat to keep the temperature of the wires below that required for the passage of flame through the meshes, and the lamp continues to burn with safety if kept in a still atmosphere.

Stephenson's lamp, on the contrary, only admits air

through a few meshes, the glass globe preventing the entry of any air or gas from the sides: therefore only a small proportion of gas can enter the interior of the

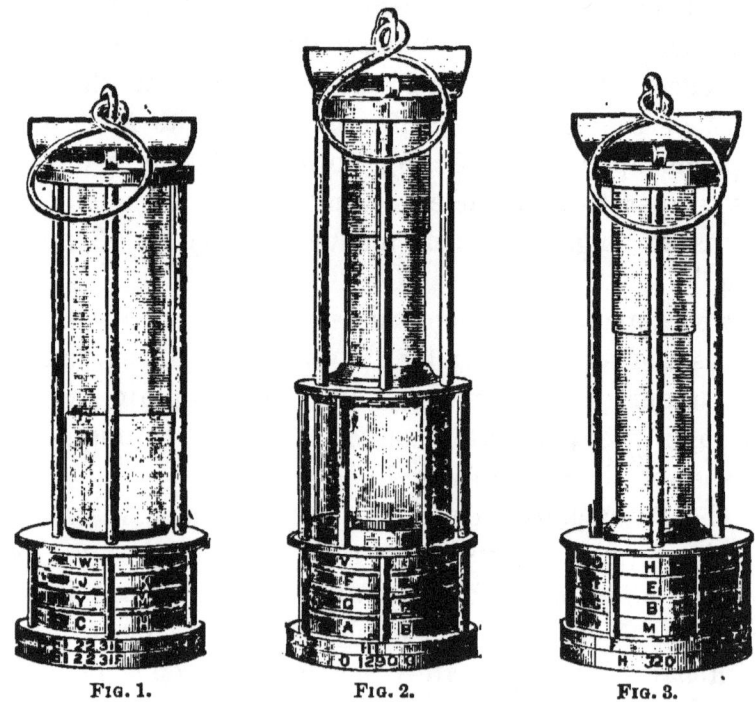

Fig. 1.   Fig. 2.   Fig. 3.

lamp: hence, never being filled with flame, the wires of the gauze remain uninjured.

Upon these principles, various modifications have been made to these lamps, until they now number a hundred or more.

§ 13. MINE VENTILATION. 35

The Clanny lamp consists of a cylinder of glass around the flame, and a wire-gauze top. A better light is produced by this combination (Fig. 2). The figures show the different lamps, with permutation-lock capable of

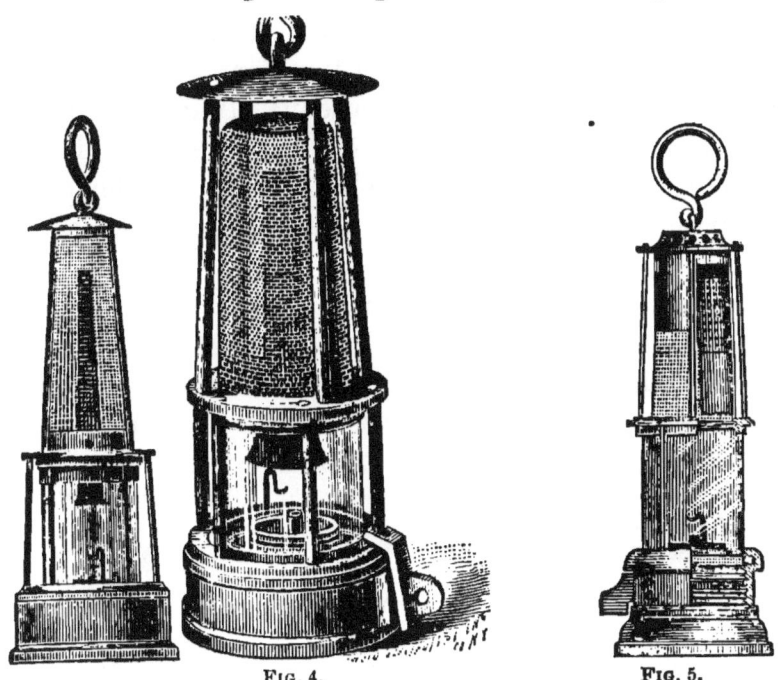

FIG. 4.   FIG. 5.

many thousand changes, so that no one but the fire-boss can unlock the lamps.

The Mueseler lamp, used in Belgium, has a glass cylinder for the light, and a gauze top (Fig. 4). There is a copper chimney to carry off the smoke of the burner, and to force the air downward between the glass cylinder and the chimney upon the flame of the burner, admitting the air through the gauze at the top.

The Boty lamp has a glass cylinder with a gauze top, but the air is admitted through a perforated copper ring at the bottom of the lamp.

The Eloin lamp has a glass cylinder, admitting air through wire gauze near the bottom of the lamp, which is thrown against the burner by a thin copper cap. No other air enters the lamp, and consequently it is easily extinguished. Many lamps are constructed to give increased light by using glass globes. The Hall lamp, with diaphoretic lens, is the most noteworthy, on account of its construction.

The Williamson double safety-lamp is a Clanny and Boty lamp combined (Fig. 5).

The illuminating power of the various lamps in most common use has been given as below; the standard being a wax candle, six to the pound:—

	Candle-power.
Davy lamp (gauze)	8
Stephenson, or Georgie	$18\frac{1}{4}$
Upton and Roberts	$24\frac{1}{4}$
Clanny (glass)	$4\frac{1}{2}$
Parish (gauze)	$2\frac{3}{4}$
Mueseler's (glass)	$3\frac{1}{2}$
Davy (without gauze)	$2\frac{1}{4}$

14. The South Shields Committee considered the Davy absolutely unsafe.

Mr. Darlington came to the same conclusion, and in answer to the question, "Is it not a fact that dust will fly off in sparks, and that one spark would create an explosion?" said, "There are very many instances of accidents taking place that we could attribute to nothing else."

Experiments made by Mr. N. Wood at Killingworth Colliery, in 1853, to ascertain at what velocity the flame may be passed in an explosive mixture of fire-damp, were as follows: —

> Davy lamp when moving 13' per second.
> Clanny lamp went out at 17' per second.
> Boty lamp passed flame when moving at 15' per second.
> Hall lamp did not pass flame at 13' per second.
> Stephenson lamp was extinguished at less than 13' per second.
> Eloin lamp went out as soon as it was filled with gas.
> Upton and Roberts lamp went out as soon as it was filled with gas.

The Belgium Commission, appointed by the king in 1868, observed, that "the Davy and Deputy lamps, when exposed for two minutes to an explosive mixture of air and lighting-gas, moving at a velocity of 4.264' per second, do not pass the flame through the gauze; but, when the velocity reaches or surpasses 7.38', the explosion is always produced on the outside, save in cases of extinction by asphyxia, caused by the admission of a large quantity of gas. It was also noticed that explosion takes place after from five to ten seconds when the velocity is 9.84′, and after from two to five seconds when the speed is 19.68′.

"With the Mueseler lamp, out of a one hundred and fifteen experiments, there were twenty-one cases of complete extinction at a velocity of 19.68′ per second.

"The Morison lamp was considered of very complicated construction, and as giving a very bad light in stagnant air. Out of eleven experiments at a velocity of 19.68′ per second, these lamps caused neither exterior explosions, nor any inflammation of gas in the exterior cylinder.

"Rapid currents of air are dangerous when their action manifests itself by the crushing of the flame upon the wicks: indeed, the relative security of the Mueseler lamp does not depend alone on the smallness of the section of the chimney at the top, or on its height, but rests essentially in the regularity of the draught."

The North of England Institute of Mining Engineers, which has been so instrumental in the furtherance of mining knowledge, appointed a committee, who rendered in their report the following concerning the velocity at which the various lamps would explode: —

§ 14 a. MINE VENTILATION. 39

	Per second.
Davy, without shield	8'
Davy, with shield	12'
Clanny	9'
Stephenson	9'
Mueseler	8'

This rather conflicts with the Belgium Report, as they claim the Mueseler lamp passes flame as easily as the Davy.

14 a. Fire-damp may be detected by the aid of a Davy or other safety-lamp. The following, taken from the Galloway Royal Society's Journal of 1876, gives the various appearances of the lamp-flame when brought in contact with air mixed with fire-damp:—

"The wick of the lamp, having been carefully trimmed, was drawn down until the flame presented the appearance of a small blue hemisphere about one-eighth of an inch high, one-quarter inch diameter at the base, and having a conical speck of yellow in the middle near the top.

"A mixture of 1 part of marsh-gas with 16 parts of air gave a voluminous waving, spindle-shaped blue cap $3\frac{3}{8}''$ high.

"1 part of marsh-gas with 18 parts of air gave a cap $2''$ high, which burned more steadily.

"1 part of marsh-gas with 20 parts of air gave a cap

$1\tfrac{5}{16}''$ high, with nearly parallel sides to about two-thirds of its height, and then tapered to a point at the top.

"1 part of marsh-gas with 25 parts of air gave a conical cap from $\tfrac{1}{2}$ to $\tfrac{5}{8}''$ high.

"1 part of marsh-gas with 30 parts of air gave a conical cap $\tfrac{3}{8}''$ high.

"1 part of marsh-gas with 40 parts of air gave a conical cap $\tfrac{1}{4}''$ high.

"1 part of marsh-gas with 50 parts of air gave a faint cap $\tfrac{1}{8}''$ high, the top having the appearance of having been broken off.

"With 1 part of marsh-gas and 60 parts of air, it was hardly possible to distinguish any thing above the small oil-flame."

CHAPTER IV.

PHYSICAL PROPERTIES OF AIR IN MOTION.

15. WIND is air in motion; and, as air is matter, it is subject to the laws which govern matter. No particle of matter possesses within itself the power of changing its existing state of motion or rest. When a body is at rest, a force is required to put it in motion; and, when once put in motion, it would continue to move on forever if a force of some kind were not opposed to it to arrest its movement. This passive property of air is called its inertia, and may be defined as opposition to change, either from motion to rest, or *vice versa*. If the air, then, had no inertia, it would not require force to give it motion, nor could it require momentum. The sailing of ships, the windmill, the tornado, are familiar examples of the power of moving air, and, consequently, proofs of its inertia. In order to pass air through a mine, certain force must be expended; and this force is what we are now about to examine. It involves a consideration of the resistances to be overcome, such as area of the airways, obstructions in the airways, and the friction against the walls or sides of the airways. When air travels the galleries of a mine, it rubs against all the exposed surfaces. This rubbing gives rise to the resistance called "friction."

Friction of air in mines is so great, that, out of every ten parts of power employed for the ventilation of a mine, about nine of the ten are used in overcoming the resistance to ventilation. As the air journeys through the mines from the downcast to the upcast, it not only meets the rubbing-surface, or sides, but often encounters short turns, brattices, etc., which adds greatly to the total resistance to be overcome.

In turning sharp corners, the air strikes square

against the face, and rebounds; thus hindering the progress of the air following, which, in turn, goes through the same operation, and hinders the air following it. It is therefore preferable to have well-rounded bends of large radius, as they produce little resistance in comparison with elbows or square bends. The consideration of the movements of air in mines or confined passages involves its density, the area, length, and perimeter of the airways, also the velocity with which the air travels.

16. To find the perimeter of an airway, we must add together the bounding-lines. The perimeter of a square airway $6' \times 6'$ is therefore $6 + 6 + 6 + 6 = 24'$. The perimeter of a circle is its circumference, and is 3.1416 times its diameter: hence the perimeter of an airway six feet in diameter is $18.8496'$. The sectional area of an airway is found by multiplying its height by its width. Thus the area of an airway $5' \times 6'$ is 30 square feet. The rubbing-surface is found by multiplying the perimeeter by the length of the airway.

Problem. — What is the rubbing-surface of an airway $500'$ long, with an area of $6' \times 6'$?

Solution. — $6 + 6 + 6 + 6 = 24'$; then $24' \times 500 = 12,000$ square feet. *Ans.*

As friction increases according to the rubbing-surface,

so will it increase or decrease according to the bounding-lines or perimeter of the airway. From this it becomes evident that the form which has the least perimeter will have the least rubbing-surface. Above, it was shown that a circular airway had less perimeter than a square airway of the same diameter; and hence, if the two airways have the same length, the circular will have less rubbing-surface.

17. In large airways the friction will be less than in smaller airways the sum of whose areas are equal to the area of the large airway.

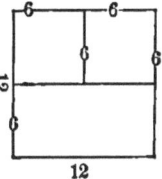

Let us suppose an airway twelve feet square in section: its area will be 144 square feet, and its perimeter $12 + 12 + 12 + 12 = 48'$.

Let us now take three smaller airways, two $6' \times 6'$ in section, and one $6' \times 12'$ in section. The aggregate areas of the three airways will be $36 + 36 + 72 = 144$ square feet: the sum of the perimeters will be $24 + 24 + 36 = 84'$. Hence we see, that, while the large airway has the same area as the aggregated smaller airways, its perimeter is much smaller than the aggregate perimeters of the lesser airways.

18. The force used to overcome the resistances

offered to the passage of air in mines is estimated in pounds to the square foot, and may be expressed as so much head of air, motive-column, or water-gauge. The motive-column has already been treated of, so our attention may be given to the water-gauge. The water-gauge is an instrument used to measure the dynamic force of a current of air. It consists of a U-shaped tube of equal area throughout. The arms are about six inches long, provided with a scale divided into inches and fractional parts of an inch, so that the difference between the height of the water in one arm of the tube and that of the other may be measured. One arm is placed in connection with the air passing in the mine, while the other is open to the air away from the mine. The difference in water-level will indicate the drag, or the resistance to the air in the mine. In some gauges, oil is substituted for water. They are made in different shapes; but the principle is the same in all. The weight of one cubic foot of water at 62° F. and 30″ barometrical pressure is 62.32102 pounds avoirdupois: $62.32102 \div 1728 = 0.036$ pound is the weight of one cubic inch of water. When the gauge measures one inch, the pressure is $0.036 \times 144 = 5.184$, or 5.2 pounds (nearly) to the square foot.

Example. — Suppose a water-gauge read 0.4″, what pressure would it indicate?

$0.036 \times 0.4 \times 144 = 2.0734$ pounds to the square foot.

§ 18. MINE VENTILATION. 45

This gauge may be used to show the force of a current produced by a fan or by a furnace, and hence is very useful as a check to the furnace-man. As it tells the amount of resistance to the air in the air-courses, their state or condition may be inferred. If the pressure per square foot exerted by the motive-column be known, the height of the motive-column may be determined.

Problem. — Suppose the temperature of the motive-column be 62° F., and the water-gauge reads 0.4″, what is the length of such motive-column?

Solution. — 100 cubic feet of air at 62°, barometer 30″, weigh 7.629 pounds. 1 cubic foot of air at 62°, barometer 30″, weighs 0.07629 pounds. The pressure per square foot as indicated by 0.4″ water-gauge is $0.036 \times 0.4 \times 144 = 2.0736$ pounds. Dividing the pressure per square foot by the weight of a cubic foot of air gives $\frac{2.0736}{0.076} = 27.28'$ as the length of the motive-column in feet.

When the height of the motive-column is known, we may find the velocity of the air in feet per second which the motive-column will produce.

Problem. — Suppose the motive-column be 27.25′ in height, what velocity per second will it produce? A body falling, acted upon by gravity, would, according to Eq. 2, attain a velocity represented by $8.02\sqrt{h}$; or substituting the value of h, which in this case is 27.25', we have $8.02\sqrt{27.25} = 42'$ per second.

TABLE IV.
WATER-GAUGE.

Water-gauge in inches.	Pressure in pounds per square foot. $P = 0.036 \times W.G. \times 144.$	Length of motive-column in feet at 62° F. $\dfrac{P}{0.0763}.$	Velocity of the air in feet per second due to motive-column. $V = 8.02\sqrt{h}.$
0.1	0.5184	6.79	20.8921
0.2	1.0368	13.58	29.5938
0.3	1.5552	20.37	36.1702
0.4	2.0736	27.16	40.7416
0.5	2.5920	33.95	45.3130
0.6	3.1104	40.74	51.1676
0.7	3.6288	47.53	55.2578
0.8	4.1472	54.32	58.5460
0.9	4.6656	61.11	60.6262
1.0	5.1840	67.90	66.0046
1.1	5.7024	74.69	69.2928
1.2	6.2228	81.48	72.3404
1.3	6.7392	88.27	75.2376
1.4	7.2576	95.06	78.1950
1.5	7.7760	101.85	80.9218
1.6	8.2944	108.64	83.5684
1.7	8.8128	115.43	86.0546
1.8	9.3312	122.22	88.6210
1.9	9.8496	129.01	90.7062
2.0	10.3680	135.80	93.1122
3.0	15.5520	203.70	114.3652
4.0	20.7260	271.60	132.1696
5.0	25.9200	339.50	147.9690
6.0	31.1040	407.40	161.8436
7.0	36.2880	475.30	174.8360

Table IV. gives the comparative height of the water-gauge and air-column at a temperature of 62° F., with pressure in pounds per square foot, and the theoretical velocity of air due to this pressure. This table, it must be remembered, is only theoretically true, on account of the enormous power required to overcome friction to the passage of air in an airway: hence, in practice, from ten to twenty times this amount of motive-column is required in order to produce the theoretical velocity.

19. "Co-efficient of friction" is a term used to represent the constant resistance met with by air during its journey through the mine. This resistance must be overcome at each point, before the air can pass that point. It varies, of course, under different conditions; but, the smoother the rubbing-surface, the less will be this resistance. This co-efficient cannot be determined with any degree of certainty except by actual experiment; and even then experimenters differ in the exact amount, because of the different conditions under which the experiments were made. Sir John Atkinson, after comparing the results of a number of experimenters, took an average between their results, and used, as the co-efficient, 0.26881 feet of air-column of the same density as the flowing air. He appears to remain in

doubt whether the mere change of temperature does or does not affect the co-efficient of resistance. As it is now taken for granted that it is not influenced by change of temperature (although it probably is), we may, for a velocity of 1,000 ~~cubic~~ feet per minute, consider the friction equal to an air-column 0.26881 feet in height, of the same density as the flowing air. This air-column is equal to a pressure of 0.0217 pound per square foot of area of section with air at 32°.

$$\frac{1.32529 \times 30}{459 + 32} \times 0.26881 = 0.0217 \text{ pound per square foot.}$$

When we consider the air-column, we may use the co-efficient 0.26881; afterwards, if we desire, we may reduce the height of the air-column thus found to pounds per square foot, as above.

We may shorten this work, however, by using 0.0217 pound per square foot of area for every square foot of rubbing-surface exposed to the air-current, at a velocity of 1,000 feet per minute, or 0.0000000217 pound for a velocity of 1 foot per minute.

We must also bear in mind, that, while we use h in the following formulas for the air-column which by its weight will produce pressure, we use 0.0217 to find the pressure, P, direct, considering all air in passing to have the same co-efficient, viz., 0.0217.

§ 20. MINE VENTILATION. 49

Mr. Thomas and several other writers consider this co-efficient as much too large: be that as it may, we can see at present no reason why it should be rejected, until the experiments already referred to approach more nearly to each other in their results. The co-efficients in practice can never be the same in different mines; but they may approach each other by making the airways as smooth, and free from obstructions, as possible.

PRESSURE.

20. Some of the formulas relating to the friction of air in mines, adopted by Mr. Atkinson in his book on "Mine Ventilation," will be retained in this treatise, so that any one studying this book will better understand his method of reasoning.

Let h = motive-column in feet.

p = pressure per square foot due to weight of h.

a = sectional area in square feet.

k = co-efficient of friction = 0.0217 pounds per square foot of area of section for a velocity of 1000 feet per minute with air at 32°. It is to be taken in the same terms or unit as p is taken in.

s = rubbing-surface.

v = the velocity of the air in thousandths of feet per minute, 1000 feet per minute being taken as the unit of velocity.

The total pressure is found by the formula

(6) $$pa = ksv^2$$

which, expressed in words, gives us the following rule: To find the total pressure due to the friction of air passing through an airway, multiply the co-efficient of friction by the rubbing-surface, and the product by the square of the velocity; or, pressure being known, multiply the pressure per square foot by the area of the airway.

21. To find the pressure per square foot,

(7) $$p = \frac{ksv^2}{a},$$

divide the total pressure by the area of the airway.

By the clearing of fractions, division, and other algebraical operations, we may find formulas which correspond to rubbing-surface and velocity, and also find the co-efficient of friction. These formulas embrace only pressure due to friction, and not that due to the creation of velocity: hence they will be more correct for long than for short airways. The symbols in these formulas are so connected with each other, that, when a sufficient number of them are known, those unknown may be found. To show the application of these formulas, suppose we have an airway $8' \times 7'$, 2,000 feet long, with the air travelling at the rate of 15 feet per second through it. What is the resistance due to friction, or the motive-column required to overcome friction in the airway?

$a = 8 \times 7 = 56$ square feet.
$k = 0.26681$.
$s = 8 + 8 + 7 + 7 = 30 \times 2000 = 60000$ square feet.
$v^2 = 15 \times 60 = 900'$ per minute, or 0.9 of 1000 feet per minute, which squared is equal 0.81.

Substituting the numerical values of these symbols in (7),
$$h = \frac{0.26881 \times 60000 \times 0.81}{56} = 233.288 \text{ feet}$$
of air-column as the pressure required to overcome the friction, and produce circulation. Taking the air at 62° F., one cubic foot, with the barometer at 30 inches, would weigh 0.0763 of a pound. If we now multiply this motive-column by the weight of a cubic foot of the air it is composed of, and divide the product by 5.2 pounds, — the pressure per square foot when the water-gauge is one inch, — we may find the water-gauge due to this pressure, and also the pressure on each square foot.

$$\frac{0.0763 \times 233.288}{5.2} = 3.42'' \text{ water-gauge.}$$

The pressure per square foot is 17.7 pounds. From this we may obtain a formula for water-gauge.

$$(8) \qquad W = \frac{\frac{ksv^2}{a}}{5.2} = \frac{p}{5.2}.$$

The same result may be obtained by multiplying the length of the motive-column by the weight of a cubic foot of air of the same density, thus obtaining the pressure direct without first finding the water-gauge.

The quantity of air passing may be found by multiplying the velocity in feet per minute by the area of airway in square feet, or

$$(9) \qquad Q = va = \frac{u}{p} = \frac{\sqrt{pa}}{ks} \times a = \sqrt[3]{\frac{u}{ks}} \times a.$$

The u in the last equation represents units of work, foot-pounds applied to circulate the air.

$$(10) \qquad u = Q \times p = vpa = HP \times 33000.$$

The HP in the last formula stands for horse-power of ventilation.

$$(11) \qquad HP = \frac{u}{33000} = \frac{Qp}{33000}.$$

When we represent the length of an airway by l, the rubbing-surface by s, the perimeter by o, we have

(12) $\qquad s = o \times l.$

(13) $\qquad l = \dfrac{s}{o}.$

(14) $\qquad o = \dfrac{s}{l}.$

CHAPTER V.

THE LAWS AFFECTING AIR IN MINES.

22. THE laws affecting the circulation of air through mines or confined passages, such as gangways, etc., have been ascertained principally by such eminent men as Magnus, Regnault, Gay-Lussac, Daubisson, Peclet, and others, and are as follows: —

1. The volume assumed by a given weight of air is inversely proportional to the pressure on each unit of surface under which it exists, so long as the temperature remains unaltered. Consequently, if we take a cubic foot of air under a pressure of five pounds, it will only be one-half a cubic foot under a pressure of ten

pounds, and one-third of a cubic foot under a pressure of fifteen pounds.

2. When the pressure is constant, the volume is uniformly increased in the ratio of $\frac{1}{459}$ part for each additional degree of heat, Fahrenheit's scale.

3. When air is discharged through orifices offering no sensible frictional resistance, the result is sixty-five per cent of the quantity due to the velocity multiplied by the area in case of a thin plate; ninety-three per cent in case of a short cylindrical tube; and ninety-five per cent when the tube is conical, and the area taken from the small end. This contraction of the flowing air, which is similar to that which takes place when water is discharged through pipes under the same conditions, has the effect of reducing the quantity discharged in a given time below that which would be due to the velocity, if it existed, over an area equal to that of the orifice or tube. This contraction of the flowing air is termed the "vena contracta."

4. When air is impelled through a confined passage, the pressure or head of air-column required for its propulsion is proportional *to the square of the velocity;* so that to double this velocity there must be four times the head; to treble it, nine times the head; etc.

(*a*) Ventilating pressure, *p*, or ("head of air," or "motive-column," reduced to pounds) total pressure,

§ 22.   MINE VENTILATION.   55

pa or *ha*, and ventilating power, *P*, are separate and distinct terms. Ventilating pressure, or simply pressure, is the force applied to each square foot of area of section to produce ventilation. That this pressure varies as the square of the velocity, as stated above, may be illustrated by the following: —

Problem. — When a mine is passing 20,000 cubic feet of air per minute with a pressure of 2.6 pounds per square foot, as indicated by 0.5-inch water-gauge, what will be the pressure if the mine pass 40,000 cubic feet per minute? This principle may be resolved simply into finding a fourth proportional; thus, —

$$v^2 : (2V)^2 :: p : ?$$
$$(20000)^2 : (40000)^2 :: p : ?$$
$$(2)^2 : (4)^2 :: 2.6 : 10.4 \text{ pounds.} \quad Ans.$$

(*b*) The total pressure is the ventilating pressure, *p*, multiplied by the area of section *a* of the airway in square feet. Thus, if the sectional area of the above airway were 64 square feet, and the pressure 0.5-inch water-gauge, the total pressure would be $pa = 64 \times 2.6 = 166.4$ pounds.

(*c*) Ventilating power is power used to obtain ventilating pressure. This power, *P*, varies as the cube of the velocity of the air-current. By this we mean, that, if we can circulate a quantity of air with 2 *HP*, we must use 8 HP if we wish to circulate a double quantity. This, as can be readily seen, is a very important factor in fiery mines; as the engine working the fan or the furnace may be called upon at any time to do double and maybe threefold duty, in case blowers or barometrical differences allow an unusual amount of gas to be given off.

Problem. — Suppose we have 20,000 cubic feet of air passing with a water-gauge of 0.5 inches, equal to a pressure of 2.6 pounds per square foot, what will be the ventilating power, P, if we double the ventilation?

Solution. — $20,000 \times 2.6 =$ units of work $= 52,000$ foot-pounds. As there are 33,000 foot-pounds in 1-horse power, we have

$$P = \frac{u}{33000} = \frac{52000}{33000} = 1.576 \; HP.$$

Again: from 22 (*a*) we find that to double the quantity we must employ four times the pressure: hence

$$P = 40000 \times 2.6 \times 4 = 416000 \text{ foot-pounds} = u$$

and

$$P = \frac{u}{33000} = \frac{416000}{33000} = 12.608 \; HP$$

or eight times the power in the first case.

5. In airways of the same sectional area, the pressure

required to propel air is proportional to the length of the passage, or, in other words, there must be double pressure for double distance.

Problem. — Suppose we have an airway 2,000 feet long, and another 4,000 feet long, of the same area; the pressure being 2.6 pounds per square foot in the shorter. What will be the pressure necessary to overcome the resistance in the longer airway?

$$2000 : 4000 :: 2.6 : ?$$
$$1 : 2 :: 2.6 : 5.2 \text{ pounds.} \quad Ans.$$

6. The pressure required to propel air through confined passages is proportional to the perimeter of the passages; the length and other data remaining constant. Thus, if we have an airway 4 feet square, and another 8 feet square, with a pressure of 2.6 pounds per square foot for the 4-foot airway, we shall have 5.2 pounds for the 8-foot airway, or

$$1 : 2 :: 2.6 : 5.2. \quad Ans.$$

7. The pressure required on each unit of surface, square inch or square foot, to propel air through a confined passage, is inversely proportional to the sectional area of the passage, when all other things are equal; so that, the greater the area exposed to the pressure, the less is the amount of pressure required for each

unit of surface. This is of great importance in ventilation, as it allows of a greater quantity of air passing with the expenditure of less power than any other means known. This principle is the one upon which the splitting of air is reasoned. In the case of the eight-foot and four-foot passages in (6), while the former required double the amount of pressure for its double-sized perimeter, that pressure would propel four times the quantity: otherwise the same expenditure of power on air in a four-foot passage would propel double the quantity it would force through a two-foot passage, or an equal quantity would be propelled by half the power. The entire mass of moving air in an airway may be considered as a column of water passing through a pipe, exposing a certain amount of surface to resistance, and hence requiring a fixed amount of pressure acting upon its sectional area to overcome such resistance when the velocity is constant: therefore, the greater the area of section, the less the amount of pressure requisite for each individual unit of such area in order to make up the gross amount of such pressure required. The steam-engine piston will require less pressure per square inch to produce a given force as the area of the piston is greater, and *vice versa*. Taking the areas given in (6), we have the following proportion:—

$$64 : 16 :: 2.6 : 0.65. \quad Ans.$$

8. The pressure required to overcome the frictional resistances encountered by air in passing through a confined passage has been found to vary with the nature of the material composing the inner surface of the airway to which the moving air is exposed in its route, as well as the mechanical state of its surface: in other words, the smoother the rubbing-surface, the less the resistance.

CHAPTER VI.

LAWS AFFECTING THE MOVEMENT OF AIR IN MINES, CONTINUED.

23. 1. In airways of the same sectional area, and which only differ in length, "the volume and velocity of air-currents are inversely proportional to the square roots of the lengths."

Problem. — When an airway $4' \times 5'$ area, and length of 4,000 feet, passes 20,000 cubic feet of air per minute, what will another airway 1,000 feet in length, with the same area and pressure, pass?

$$\sqrt{1000} : \sqrt{4000} :: 20000 : x$$
$$\sqrt{1} : \sqrt{4} :: 20000 : 40000 \text{ cubic feet.} \quad Ans.$$

Again: the velocity in the second airway will be

$$\sqrt{1000} : \sqrt{4000} :: \frac{20000}{20} : x$$

that is

$$\sqrt{1} : \sqrt{4} :: 1000 : x,$$

or

$$1 : 2 :: 1000 : 2000'\text{ per minute. }Ans.$$

That is, the volume and velocity in the 1,000-foot airway are twice the volume and velocity in the 4,000-foot airway.

2. The volume passing through airways of similar form but unequal size will be greater as the area of section is greater, other data being the same; or, the pressure and other data remaining constant, the quantity will be directly proportional to the areas: —

$$a : A :: v : V.$$

3. The volume passing, when areas and other data are equal, is inversely proportional to the perimeter of the sections of the airways. The circle passes the greater quantity of two airways, one of which is a square with a side equal to the diameter of the circle. (See § 16.)

4. The quantity or volume of air passing through an airway varies as the square root of the rubbing-surface.

5. Friction diminishes in proportion to the square root of the velocity, and increases according to the square of the velocity; i.e., friction varies as the square of the velocity.

6. Pressures are proportional to the squares, and the powers are proportional to the cubes, of the quantities of air passing through airways.

7. The quantity of air passing through airways of different areas, other things being equal, is according to the square root of the area multiplied by the area.

PROBLEMS TO ILLUSTRATE THE FOREGOING LAWS.

24. (*a*) If 20,000 cubic feet of air can be produced in an airway of 60 feet area with a certain pressure, how much air will the same pressure produce in an airway of 30 feet area?

Assume the perimeters to be 32 feet for 60-foot airway, and 22 feet for 30-foot airway. Had the relationship which existed between the perimeter and the area of the larger airway been maintained in the smaller airway, then the quantity of air flowing through the latter would be directly proportional to its area (2), pressure, etc., remaining constant. That is to say, the ratio of the area to the perimeter of the larger airway is as 60 to 32. In the smaller airway we have one-half the area; and, had the smaller perimeter been one-half the larger, then the ratio would have been unaltered, and the quantity of air would have been one-half also (3). But, instead of the smaller perimeter being 16 feet, it is 22 feet. Now, the problem resolves itself simply into finding a fourth proportional; and, bearing in mind (4) that the quantity varies inversely as the square root of the rubbing-surface, we have,

$$\sqrt{22} : \sqrt{16} :: 10000 \text{ cubic feet} : x$$
$$\therefore x = 8528 \text{ cubic feet per minute.} \quad Ans.$$

In the above calculation the length of the airway is taken as unity; because any given length will be a factor common to both terms of the first ratio, and hence may be eliminated.

To prove this method of reasoning, we may work it out according to Atkinson's method:—

Taking the length of both airways as 1,000 feet, we have, for the larger,

$$s = 32000 \text{ square feet.}$$
$$v = \frac{20000}{60} = 0.333$$

in thousandths of feet per minute. Substituting in formula (7)

$$h = \frac{ksv^2}{a} = \frac{0.26881 \times 32000 \times (0.3 \times 0.3)}{60}$$
$$= 15.9295' \text{ of air-column.}$$

§ 24.   MINE VENTILATION.   63

For the smaller airway,
$$s = 20000,$$
and
$$v^2 = \frac{15.9295 \times 30}{0.26881 \times 20000} = \frac{477.885}{5913.82} = 0.0808082$$
$$\therefore v = \sqrt{0.0808082} \times 1000 = 284.27 \text{ feet per minute};$$

and, as the quantity passing is found by (9), § 21, we have $284.27' \times 30 = 8528$ cubic feet per minute.

Problem (b). — Suppose, now, we take two airways whose lengths are each 35 units, and which have the same area; the quantity of air passing through each being 9,000 cubic feet. What will each circulate of the total amount, if their lengths be in the ratio of 3 to 4?

Solution. — From (1) and (4), § 23, we deduce two proportions, — one for the longer, and one for the shorter airway. Now, if these airways were in the ratio of 3 to 4 in length, then one would be 30 units and the other 40 units in length, and the quantities of air which would flow through them under the same conditions may be computed thus: —

For the longer airway
$$\sqrt{40} : \sqrt{35} :: 9000 : x$$
or
$$\sqrt{8} : \sqrt{7} :: 9000 : x$$
and
$$8 : \sqrt{56} :: 9000 : 8418. \quad Ans.$$

For the shorter airway

$$\sqrt{30} : \sqrt{35} :: 9000 : x, \text{ or } \sqrt{6} : \sqrt{7} :: 9000 : x$$

and hence

$$6 : \sqrt{42} :: 9000 : 9721 \text{ cubic feet. } Ans.$$

Total,

$$8418 + 9721 = 18139 \text{ cubic feet.}$$

Proof,

$$\sqrt{3} : \sqrt{4} :: 8418 : 9721$$
$$\sqrt{4} : \sqrt{3} :: 9721 : 8418.$$

(*c*) Suppose we have two airways of the same sectional areas and lengths, each passing 9,000 cubic feet as before. Suppose each to have 4 units of lengths. If one of these airways be shortened to unity, it will have but ¼ the rubbing-surface of its former length; and the volume of air will be, according to (1), found thus:—

$$\sqrt{1} : \sqrt{4} :: 9000 : 18000.$$

Again: let the length of one of the airways be increased fourfold, then the volume of air will have four times the rubbing-surface, and by (4) we have

$$\sqrt{16} : \sqrt{4} :: 9000 : 4500.$$

(The above illustrates in a striking manner the effect that rubbing-surface has of diminishing the flow of air through a gallery.)

§ 24. MINE VENTILATION. 65

The total volume will be 22,500 cubic feet; and, notwithstanding the rubbing-surface is more than doubled, the volume is only increased by twenty-five per cent. The pressure required to circulate the air under each set of conditions is precisely the same; for the smaller rubbing-surface multiplied by the square of the highest velocity is equal to the greater rubbing-surface multiplied by the square of the lower velocity; thus $\left(\frac{9000}{40}\right)^2 \times 4 = 202500$, and $\left(\frac{4500}{40}\right)^2 \times 16 = 202500$, and $\left(\frac{18000}{40}\right)^2 \times 1 = 202500$, if we assume the areas of the above to be 40 square feet.

That the pressure remains the same may be shown by Atkinson's method: —

Let the above unit of length be assumed as 100 feet, then equal airways are 400 feet in length; and, if we assume the area and perimeter to be respectively 25 and 20 feet, we may find the pressure, thus: —

(1) $h = \dfrac{0.26881 \times 400 \times 20 \times 0.1296}{25} = 11.148$ ft. head.

(2) $h = \dfrac{0.26881 \times 100 \times 20 \times 0.5184}{25} = 11.148$ ft. head.

(3) $h = \dfrac{0.26881 \times 1600 \times 20 \times 0.0324}{25} = 11.148$ ft. head.

(*d*) If, through two airways 6 feet square, 9,000 cubic feet of air flow per minute, and one of them be altered in section to a circle (its area being unaltered), then the quantity of air circulating may be computed, thus:—

$$\sqrt{21.27} : \sqrt{24} :: 9000 : 9560. \quad Ans.$$

The volume flowing through both airways would now be 18,560 cubic feet; but, in the event of this quantity being reduced to 18,000 cubic feet, the circular airway would have more flowing through it than the other.

(*e*) What volume of air would flow through an airway 5 feet square, if 6,000 cubic feet flow through an airway 10 feet square, the pressure and length being the same?

Solution. — The volume varies as the square root of the rubbing-surface (§ 23, 4), and directly as the area (§ 23, 2): hence we have a compound proportion,

$$\begin{matrix}\sqrt{40} : \sqrt{20} \\ 100 : 25\end{matrix} :: 6000 : x$$

or

$$x = \frac{\sqrt{20} \times 25 \times 6000}{\sqrt{40} \times 100} = 1061 \text{ cubic feet.} \quad Ans.$$

It may be reasoned thus: as the area was reduced one-fourth, the resulting volume would be one-fourth also, namely, 1,500 cubic feet; but, instead of the perimeter being reduced to one-fourth, that is, 10 feet, it is only reduced to 20 feet. The volume may now be found, thus: —

$$\sqrt{20} : \sqrt{10} :: 1500 : x$$

or

$$\sqrt{2} : \sqrt{1} :: 1500 : 1061 \text{ cubic feet. } Ans.$$

(*h*) Suppose we have a pressure equivalent to 40 *HP*, giving a circulation of 120,000 cubic feet per minute: what quantity will a pressure equivalent to 32 *HP* give? From § 23, 6, we have

$$\sqrt[3]{40} : \sqrt[3]{32} :: 120000 : 111398 \text{ cu. ft. per minute. } Ans.$$

Proof. $(111398)^3 : (120000)^3 :: 32 : 40.$ *Ans.*

(*i*) Suppose we have a pressure equivalent to 32 pounds per square foot, circulating 107,350 cubic feet per minute, what pressure will circulate 120,000 cubic feet?

Solution. $(107350)^2 : (120000)^2 :: 32 : 40.$ *Ans.*
Proof. $\sqrt{40} : \sqrt{32} :: 120000 : 107350.$ *Ans.*

25. From the proof of the last problem we see that air may be measured by the pressure, or, what amounts to the same, the water-gauge, and we can say the quantity of air passing in a mine is according to the square root of the water-gauge.

Problem.—Suppose we have 30,000 cubic feet of air passing when the water-gauge is 1.6 inches, what quantity will pass with 2.5 inches of water-gauge?

$$\sqrt{1.6} : \sqrt{2.5} :: 30000 : x$$

or

1.2649 : 1.5811 :: 30000 : 37500 (nearly) cubic feet.

TABLE V.

SQUARE ROOT OF WATER-GAUGE.

W.G.	$\sqrt{W.G.}$	W.G.	$\sqrt{W.G.}$	W.G.	$\sqrt{W.G.}$
0.1	0.3162	1.2	1.0954	2.3	1.5165
0.2	0.4474	1.3	1.1401	2.4	1.5491
0.3	0.5477	1.4	1.1832	2.5	1.5811
0.4	0.6324	1.5	1.2247	2.6	1.6144
0.5	0.7071	1.6	1.2649	2.7	1.6431
0.6	0.7745	1.7	1.3038	2.8	1.6733
0.7	0.8366	1.8	1.3416	2.9	1.7029
0.8	0.8944	1.9	1.3784	3.0	1.7320
0.9	0.9486	2.0	1.4142	3.5	1.8460
1.0	1.0000	2.1	1.4491	4.0	2.0000
1.1	1.0488	2.2	1.4832	4.5	2.1213

CHAPTER VII.

VENTILATION OF SINGLE PITS OR DRIFTS.

26. DURING the sinking of shafts, or driving of drifts, the ventilation is caused by the diffusion of gases, and by the temperature of the atmosphere or outer air and that of the mine. It can hence only be temporary, becoming insufficient as soon as certain variable and often very small depth is attained, unless recourse be had to some other means. The simplest mode of attaining this end consists in dividing the pit by a wooden brattice, so that the warm air may ascend one side, while the cool air will descend the other. If this does not produce circulation, it can generally be attained by connecting one side of the brattice with a high chimney, which, by establishing a difference between the levels of the orifices, causes a natural circulation of air. If compressed air or steam be used for drills, or if pumps be suspended in the pit, the chimney will be unnecessary. A fire may be placed at the bottom of the pit to warm the air on one side the brattice, or a small hand centrifugal fan be employed to advantage. In drifts, the brattice is placed parallel to the axis, and disposed in a horizontal or vertical

plane as better suits the case. When placed in the vertical plane, one of the compartments may be used as a tramway and intake for the air: the other, communicating with the upcast pit, is used entirely as a return. When a horizontal brattice is used, it may be at the feet or head of the workmen.

In this case the brattice is used to facilitate the ventilation of drifts, and also the driving and sinking of internal drifts, by putting the airway one side of the brattice in communication with the downcast. Within the last fifteen years, miners have superseded this method, wherever practicable, by "brattice-cloth" nailed to a series of props. The cloth is made impermeable to the air by being covered with tar, or dipped in a solution of soda silicate. Treated in the latter manner, it is rendered incombustible, as well as impermeable to the air. Brattice-cloth is only used to secure temporary ventilation, and should never be permitted, except at the face, or where it is subjected to constant scrutiny, as it may decay; and inattention to its repair or renewal might cause serious accidents. Another method of conducting air into drifts is by a continuous wooden box. This box may be placed in any position where it will be out of the way, with one of its extremities connected with a chimney, if it be driven from the day; or with the return airway, if the sinking or drift be made in the mine. Air-boxes, whose use is consistent with economy, if of a great length, will not supply sufficient volume of air, owing to the great absorption of the "motive-column" by friction. Tubes are made of other materials than wood, such as zinc, canvas, and paper; but they have never come into great demand. These air-boxes or tubes may be used to convey fresh air to the face of the workings, or return the foul air. It is preferable to employ them for the first method, rather than for the second; for more air is thrown upon the face of the workings, because, when suction is employed, the air will leak through the box before it reaches the mouth of the box at the upcast, also the contraction of the air-vein at the entry of the box must be greater in the suction than in the forcing principle.

Again: the air will be purer when delivered to the face fresh from the day, for by suction it must necessarily bring to the face gases and foul air met with along its route. It is also evident that the diffusion of fire-damp by the mechanical action of the air will be better by the forcing-system. When the excavation has attained the extreme point where air-boxes and brattice act with sufficiency, we must replace these modes of ventilation (purely temporary) by other more perfect means, and, above all, not employ air-boxes in order to drive winning-places, because the ventilation would be insufficient, owing to the length which would be required; and impure air would injure the miner's health.

Again: should an explosion take place, the boxes would be destroyed, ventilation stopped, and the afterdamp would spread through the workings, and cut off all chance of escape to the uninjured, and all hope of relief to the injured. As soon as the shaft wins the coal, it should be connected with a neighboring shaft, so as to produce a regular and continuous circulation of air. This is usually effected by driving two ways, forming a single conduit with two sides of the pit, so that the air must pass around the face of the drift. Sometimes a single drift is driven, divided by a brattice; sometimes two parallel drifts, separated by a wall of coal, cut through or holed at equal distances by cross-drifts or stentings, which are afterwards closed up, only leaving the one nearest the face open. After communication is made, the brattice should be removed. *

MINES WITH TWO ORIFICES.

27. A colliery, which, over a small area, has many openings to the day of great sectional area but of little length, requires little or no mechanical ventilation. When the pits are less numerous, and the galleries are longer, we should exercise great care in the distribution of the air-current. Formerly all the galleries were united so as to form one long and sinuous airway, of which each extremity emptied into the atmosphere directly, or into the pits. The air descended one of these, traversing its whole length, and finally ascended the upcast shaft. Places were stowed, and doors erected, so as to direct the air-current at various places. The inconveniences of this system arose from the length of the air-course, which became so much more as the mine developed, and which by the friction diminished the volume of air. The current in some part of the mine is pretty sure to meet with gases of some nature or other, so that, if they be mephitic, the air will pass over the workmen in the return, and injure their health; and, if the air has acquired sufficient inflammable gas, the same workmen may explode it with their lamps. In two collieries alone, where this system of ventilation was adopted (Lundhill and Risca), three hundred and twenty-nine lives were lost. The tendency of a current of air is naturally to choose, in passing from point to point, the nearest route, and that which has the largest section. If allowed to pursue its own course, the air will not always be renewed in galleries which require it.

28. To overcome this difficulty, doors and regulators are employed; the former to conduct the air where it is needed, the latter to cause a diminution of the volume of air in certain portions of the mine where only a portion of the current is needed. Should permanent stoppings be required for old ways or abandoned passages, they may consist of solid masonry, tightly stowed, so as to leave no passage for a single thread of air: on the contrary, they must be movable if the trams and men pass along the gallery. For this purpose, recourse is had to doors, which are usually placed in pairs, but so far apart that one may be shut before the other is opened. Regulators are employed to prevent more than a required quantity of air from passing into certain districts, and to cause the excess to pass into other parts of the mine. They are simply sliding-doors, whose open area may be increased or diminished as required by circumstances, and which can be locked fast at any area, the key, of course, being retained by the boss.

(*a*) To illustrate these regulators, let us assume an airway 5′ × 5′ passing 10,000 cubic feet of air per minute: a regulator is put in, which contracts it to 20-foot area. Find the quantity of air that will pass, power and pressure remaining the same.

From § 23, 5, we find that "the resistance increases as the square of the velocity of the air-current," also we know that sudden contraction of an airway will cause diminution of the discharge; then, according to § 23, 7, we have

$\sqrt{25} \times 25 : \sqrt{20} \times 20 :: 10000 : 7152$ cubic feet. *Ans.*

(*b*) As we have regulated the discharge, and diminished the quantity, we may increase the discharge by enlarging the airway.

Problem. — What volume of air would flow through an airway 10 feet square, when 6,000 cubic feet per minute flow through an airway 5 feet square, the pressure and length being the same?

$\dfrac{\sqrt{20} : \sqrt{40}}{25 : 100} :: 6000 : x$, or $\dfrac{\sqrt{40} \times 100 \times 6000}{\sqrt{20} \times 25} = 33840$. *Ans.*

In the above case the area was enlarged four times. The resulting volume would have been increased four times also, viz., 24,000 cubic feet, had it not been that the perimeter only increased to 40 feet instead of four times its former size, or 80 feet. The volume may be found thus: —

$\sqrt{80} : \sqrt{40} :: 24000 : 33840$ cubic feet. *Ans.*

CHAPTER VIII.

SPLITTING THE AIR-CURRENT. — BUDDLE'S METHOD.

29. THE splitting of the air-current of a mine, when not carried to extremes, is very advantageous, as it secures a greater volume of air at the expense of the same motive-power. The different divisions of a mine are never equally developed; and, if proper care be not taken, districts requiring the most air will receive only a small portion, while other districts requiring but little will receive large quantities. Common sense suggests, that, of two developed districts of a mine, the more developed of the two should receive the larger quantity of air: the air should therefore be distributed in a systematic manner, basing the quantities for each district on to the number of bends, extent of rubbing-surface, and other resistances. This aim is attained, as mentioned above, by the "sliding shutter," which, being placed so as to admit a small volume of air, causes whatever excess there may be to pass into other parts of the mine where the air is more needed.

It may happen that a "blower" of gas is met with, which renders the air explosive in its district. When it is perceived, the regulators should be immediately opened, and a greater circulation produced, in order to carry away the inflammable gas. During this time work in the other districts need not necessarily be stopped, unless it is found necessary to pass the whole current through the dangerous district.

If the gases be given off in such quantities that an explosion takes place before the miner perceives the explosive state of the air, the workmen in the other districts will be protected from the flame of the explosion, and the "after-damp" will only momentarily check the current; and, if the return airways are of large sectional area and no great length, the current will scarcely be interrupted, save, perhaps, in the district where the explosion occurred, the doors and regulators of which have been destroyed. Where gas is given off at several points in a mine, ventilated by a single current, the aggregate amount may be sufficient to render the current explosive, and cause an accident.

30. Either to increase the ventilating pressure, or to lessen the extent of rubbing-surface exposed to the air circulating in mines, is a very slow and costly manner of increasing the amount of ventilation; but by judiciously dividing the current, and leading it into several airways, the circulation will be much more active than if only one continuous current were used. Let us, as an illustration, assume a mine divided into four equally developed districts of the same area; and also the same mine arranged so as to form one continuous airway, whose length will be four times greater than that of one of the districts. Let $a =$ area, $p =$ perimeter, $l =$ length, and $Q =$ quantity of air circulating; then we have, as the value of the friction of the air,

$$\frac{plv^2}{a} = \frac{pl}{a} \times \frac{Q^2}{a^2}$$

multiplied by a co-efficient to be determined by experiment. In the first case, the airways may be considered as a single airway, with a length l, a section $4a$, and a perimeter $4p$: the equation then becomes

(a) $\qquad \dfrac{4pl}{4a} \times \dfrac{Q^2}{16a^2}.$

Secondly, the area and perimeter being a and p, the length will be $4l$, whose "drag" or resistances due to friction, are expressed by

(b) $\qquad \dfrac{p4l}{a} \times \dfrac{Q^2}{a^2}.$

On dividing equation (a) by (b) we find (b) is $\frac{1}{64}$ of equation (a), i.e., the friction of (a) is $\frac{1}{64}$ that of (b); or we may say "that the resistance to the motion of the air in a mine requires in a single current a motive-force sixty-four times greater than when the air is divided into four currents of the same sectional area and aggregate length.

The same is true for bends and contractions that require a motive-force equal to the square of the volume of air circulating in each district. Thus, in the first case, the volume of air was $\frac{Q}{4}$, the motive-force required to overcome this resistance $m \times \frac{Q^2}{16}$; m being a numerical co-efficient depending upon the number of bends or strictures.

In the second case, the volume Q must be multiplied by a co-efficient, which, owing to the quadruple length of the airway, must be equal to the sum of the four co-efficients of each district, or $4mQ^2$, a value sixty-four times the first.

31. By splitting the air-current, or leading it off from the downcast into the different districts to be ventilated, we have two decided advantages over the old method. In the first place, we obtain purer air for each district; for each split takes its fresh air directly from the downcast.

In the next place, we obtain more air by decreasing the velocity. As the friction increases as the square of the velocity, we lessen the friction by lessening the velocity, and using the same power: we also decrease the cost of ventilation more than by any known means as yet. Were it not for the resistances of the shafts, the results of splitting the air would be easily calculated.

Problem. — To show the effect of splitting, without considering the shaft resistances: suppose we have a mine circulating 18,000 cubic feet per minute, and that the area of the gallery is 25 feet, the length 1,000 feet. What amount of air will circulate when the current is split into two, three, four, and five equal divisions, the pressure remaining constant?

The effect of splitting into two, three, four, and five equal air-courses will be to double, treble, etc., the areas without altering the rubbing-surface, because the area after splitting is two, three, etc., times that of the original airway, although the rubbing-surface remains the same; and as the quantity may be found by the formula

(x) $$q = \sqrt{\frac{pa}{ks}} \times a$$

we may substitute the values of the symbols, and find the quantities of the various splits under consideration.

Since, however, in the above formula, p, k, and s remain constant, and a varies, we may simplify the work by eliminating them, and use $\sqrt{a} \times a = q$, which corresponds to one of our former laws; viz., "that the relative quantities will be according to the square root of the area multiplied by the area, and then multiplied by the original quantity flowing through the mine." The pressure may be found thus:—

$$p = \frac{ksv^2}{u} = \frac{0.0217 \times 20000 \times (0.72)^2}{25} = 8.999 \text{ pounds.}$$

Using now formula (x)

$$q = \sqrt{\frac{9 \times 50}{0.0000000217 \times 20000}} \times 50 = 50910$$

$$q = \sqrt{\frac{9 \times 75}{0.0000000217 \times 20000}} \times 75 = 93525$$

$$q = \sqrt{\frac{9 \times 100}{0.0000000217 \times 20000}} \times 100 = 144000$$

$$q = \sqrt{\frac{9 \times 125}{0.0000000217 \times 20000}} \times 125 = 200070.$$

The question may be worked, without reference to the actual dimensions of the areas and rubbing-surfaces, thus:—

$$\sqrt{1} \times 1 : \sqrt{2} \times 2 :: 18000 : 50910$$
$$\sqrt{1} \times 1 : \sqrt{3} \times 3 :: 18000 : 93528$$
$$\sqrt{1} \times 1 : \sqrt{4} \times 4 :: 18000 : 144000$$
$$\sqrt{1} \times 1 : \sqrt{5} \times 5 :: 18000 : 200070.$$

The difference between the two calculations is owing to the treatment of the decimals, and assuming the pressure more than it really is.

(*b*) There is a difference between splitting the air, and adding an air-course of the same length and area. In the above example the area was doubled while the length remained unchanged; but, if we were to add an air-course of the same length and area, we would double the rubbing-surface and the area. In fact, we cannot call such an arrangement a split; for, if both received an equal volume of air, they would both have the same pressure; but that pressure would only be a small part of what it was in the above example, first case, on account of the increased area: therefore, instead of increasing the quantity of air, we diminish it.

To illustrate this let $a = 25$ feet, $s = 24{,}000$ feet in an airway passing 15,000 cubic feet per minute: add an airway of the same length and area, what quantity will flow through each? *Ans.* 22,900 cubic feet.

The units of power necessary to circulate the above quantity may be found by multiplying the pressure by the quantity of air circulating per minute.

$$\text{(z)} \qquad u = \frac{ksv^2}{a} \times Q = p \times Q.$$

With an additional airway of the same length, a and s will be doubled; and, as the velocity decreases as the cube root of the power, we have the formula

$$V = \sqrt[3]{\frac{u}{ks}} \times a$$

to obtain the quantity of air in cubic feet circulating per minute.

(c) When we have splits of different lengths, and wish to know the size of regulators so as to allow the same quantity of air to each division, we may find them when we know the size of the first regulator.

Problem. — Suppose we have five different splits in a mine, 200, 300, 400, 700, and 800 yards long respectively, and the regulator placed at the entrance of the 200 yards' airway be 9 square feet. What will be the area of the other regulators so as to allow the same quantity of air to each split?

$$\sqrt{300} \div \sqrt{200} = 1.22404.$$
$$\sqrt{400} \div \sqrt{200} = 1.41421.$$
$$\sqrt{700} \div \sqrt{200} = 1.87083.$$
$$\sqrt{800} \div \sqrt{200} = 2.00000.$$

This gives us the rate at which the friction increases when compared with the first regulator; and hence, if we multiply each of the above quotients by 9 respectively, it will give us the area of the required regulators, thus: —

$$1.22404 \times 9 = 11.0'$$
$$1.41421 \times 9 = 12.7'$$
$$1.87083 \times 9 = 16.8'$$
$$2.00000 \times 9 = 18'.$$

(*d*) Again: if we wish to divide a given quantity of air so that it will be distributed, as it is needed, into unequal quantities in different divisions of the mine, we may do so in the following simple manner.

Problem. — Suppose we have 50,000 cubic feet of air to be distributed in five divisions, each to obtain respectively 15,000, 12,000, 10,000, 7,000, and 6,000 cubic feet per minute, to travel at a velocity of 5 feet per second; which velocity is sufficient to render any discharge of fire-damp harmless, unless it happens to be a very exceptional case.

Taking the first case as an example, 15,000 cubic feet to be distributed at a velocity of 5 feet per second, $5' \times 60 = 300'$ per minute, $\dfrac{15000}{300} = 50$ feet area, and $\sqrt{50} = 7.07$ feet, size of the regulator necessary for the first split. Proceeding in the same manner, the regulators to admit the above quantities of air may be found.

32. Airways are seldom of the same area in the same mine, but are subject to the same pressure in ventilation; and, as each airway takes up its part of the pressure to overcome the resistances which the air encounters while passing through it, we may find the pressure necessary to overcome the resistances of the whole mine by adding the several pressures, or the pressure for each airway, together. If a mine be ventilated by a number of airways of different areas and lengths, they must all be considered as subject to one common pressure, and the quantity of air passing in each may be found by the formula,

$$\sqrt{\frac{a}{s}} \times a, \text{ or } \sqrt{\frac{a^3}{s}}.$$

This is derived from the formula $q = \sqrt{\frac{pa}{ks}} \times a$, in which k and p are common factors for each airway, and hence need not be considered.

Problem. — Suppose we have three airways, A, B, and C. A has an area of 30 square feet, and a rubbing-surface of 66,000 feet. B has an area of 36 square feet, and a rubbing-surface of 96,000 feet. C has an area of 25 square feet, and a rubbing-surface of 40,000 feet. What quantity of air will pass along each, if the total quantity passing be 50,000 cubic feet per minute?

For the sake of brevity, the rubbing-surfaces may be reduced to the lowest whole numbers, and still remain in the same proportion to each other by dividing by 2,000; then

$$A = \sqrt{\tfrac{30}{33}} \times 30 = 28.602$$
$$B = \sqrt{\tfrac{36}{48}} \times 36 = 31.176$$
$$C = \sqrt{\tfrac{25}{20}} \times 25 = 27.950$$
$$\text{Total} \ldots 87.728$$

The proportional part passing in each airway may now be found from the simple proportions: —

$A = 87.728 : 28.602 :: 50000 : 16301.5$ cubic feet.
$B = 87.728 : 31.176 :: 50000 : 17768.5$ cubic feet.
$C = 87.728 : 27.950 :: 50000 : 15930.0$ cubic feet.

33. Find the quantity of air which will pass through a mine of the dimensions as given in the following table, with a total pressure of six pounds per square foot. In this case, the upcast and downcast shafts are given. A, B, C, and D are splits, subject to the same pressure. We may assume any quantity of air we please — say,

§ 33.   MINE VENTILATION.   87

50,000 cubic feet — to pass through each division, and then find the actual quantity which will pass under six pounds' pressure.

Before we can do this, however, we must calculate the pressure for the circulation of 50,000 cubic feet; then, by adding the several pressures together, place them in proportion, as shown in column VIII. of the table.

TABLE VI.

	I.	II.	III.	IV.	V.	VI.	VII.	VIII.	IX.
	Size.	Area.	Perimeter.	Rubbing-surface.	$\sqrt{\frac{a}{s}} \times a$.	Deduced from v. Assumed quantity, § 32.	Pressure for VI. according to $p = \frac{kv^2}{a}$	Actual pressure derived from VII. 8.328:6::2.470:1.78. *Ans.*	Actual quantity derived from VIII. by $q = \sqrt{\frac{pa}{ks}} \times a$
Upcast	9×8	72	34	17000	. .	50000	2.470	1.78	42400
Downcast	8×8	64	32	16000	. .	50000	3.316	2.40	42400
A	7×7	49	28	42000	74.848	18115.5			15412
B	8×4	32	24	36000	42.666	10326.5	2.542	1.82	8737
C	5×6	30	22	22000	49.543	11391			10146
D	5×5	25	20	20000	39.528	9567			8105

Next, by the use of the formula $q = \sqrt{\frac{pa}{ks}} \times a$, we may obtain the actual quantities. Columns VI. and VII. are obtained as shown by § 32. The pressure passing 50,000 cubic feet, we find aggregates 8.328 pounds; then

$$\sqrt{8.328} : \sqrt{6} :: 50000 : 42400 \text{ cubic feet}$$

the actual quantity, as is shown by the table.

CHAPTER IX.

34. As yet no rule has been established as to the quantity of air necessary for ventilating mines of different capacities: consequently, sometimes as much air is sent into a mine employing two hundred persons as there is into one in which twice that number are constantly engaged.

Scientifically ventilated mines contain a certain allowance of air for each person working under ground, as well as for each light and horse; also for various other purposes. Some persons think they can determine in a general manner the volume of air necessary for a mine, basing it upon the following single element, the number of miners employed, and giving to them a greater or less volume of air, varying with the presence or absence of fire-damp. Let us see if this be correct, and, to do so, assume that each workingman, exclusive of horses and lights, requires one hundred cubic feet per minute, taking as examples two mines equally developed, although, by reason of the nature of the coal-seam and the mode of working, the tonnage hewn by each man may be different, and only require sixty men in the first mine, whilst the second requires a hundred and fifty men.

The first, we will say, is from necessity wrought in two workings $6' \times 5'$. The volume per man being 100 cubic feet, the total volume required is 6,000 cubic feet per minute, which, divided by the area, $30 + 30 = 60$ square feet, gives 100 feet per minute as the velocity of the air. In the second case, one airway is required with an area of $8' \times 5' = 40$ square feet. The total volume circulating will be $100 \times 150 = 15,000$. cubic feet per minute, and the velocity will be 377.5 feet per minute. The first velocity will be insufficient to sweep away the gas, and prevent the heating of the air; while the other is too great, and would incommode the workmen. Thus a volume of air exclusively founded on the number of men employed is incomplete. The extreme limits of velocity — which largely regulate the temperature, and indirectly the capacity of the air for water-vapor — must be fixed. To do this, a record of the development of the working; the method of working; the nature of the seam; the splits, if any, or, if not, the main current; and the more or less abundant production of gases, must be kept, as it is impossible to fix in a lump otherwise the volume of air necessary for a fiery mine, where the evolution varies in such great limits. On account of the great complication of such calculations, no general system can be established. "It appears, under the circumstances, to be indispensable to proceed by analogy, by classing mines according to the more or less favorable conditions for ventilation in which they are situated, and to form groups, to the members of which a single and absolute rule can be applied."

In some mines, of course, more air is required than in others; but, for sanitary purposes alone, 120 feet per minute is quoted as the minimum for each man and boy; but, where gas is given off, twice that amount should be allowed. No person should be allowed to work in a stagnant atmosphere, while the working-places and goaves where the gases congregate should have a supply of air large enough to dilute and deprive them of their power. In all excavations where air is renewed, and in the galleries of mines in particular, carbonic-acid gas is continually found in more or less quantity. The ventilation should be sufficient to draw it constantly away, and to keep that quantity which is mixed with the air beneath that limit beyond which it would become injurious to the workmen.

Mr. Richardson, who paid a great deal of attention to the subject, estimated that the quantity of air required for vital chemical purposes for each person was upwards of 1,000 cubic feet per hour. Of this, 84 cubic feet were for the breathing of each person; 62.8 cubic feet, for displacing carbonic acid; 258.4 cubic feet, for diluting nitrogen; and 27 cubic feet, for displacing perspiration. In addition, 59.3 cubic feet should be allowed for the combustion of each light, and 2585 cubic feet for one horse. This is not considered an extravagant estimate; and some hold that it is not enough for diluting all the gases, nor for removing the air after being breathed.

"By some modes of ventilation, there are contrivances for enabling the men to breathe over and over again the same air, and so accumulate nuisances; and this is more especially the case in mines which do not give off fiery gases." Such things, however, should not be tolerated at the present age in any district. There is now no difficulty in providing air in sufficient quantities to dilute the gases given off in fiery mines, and so render them harmless. Where the furnace is used for ventilation, it has been calculated that the cost of ventilating the most difficult mines, and where there is a large escape of gas, need not exceed an English penny per day per man, and not half so much in mines where little or no gas is given off. The cost of fan-ventilation is not as great as the furnace, although the first cost is larger. As before stated, enlargement of airways, and judicious division of the current into several splits (which should begin as near as possible to the downcast), will bring the air much purer and cooler to the miners, and also greatly increase the ventilation.

PREVENTION OF COLLIERY EXPLOSIONS.

35. Much has been written and said upon this subject, yet every now and then the public is startled with news of a recent explosion where numbers of lives have been lost. The investigations which follow are not always satisfactory, either to the families of the deceased or the managers. Instances may be cited where the injured have received the blame for the negligences of either the manager or his " fire-boss."

There are laws in some of the States which require the managers to allow each miner a certain quantity of air.[1] Whenever there is " bad air," the miner should inform the " mine-boss," and, in case the evil is not speedily remedied, report the fact to the " mine-inspector." Miners should not neglect to do this, even if it costs them " their job;" because they risk the lives of

---

[1] See Sect. 7, Pennsylvania Mine Laws.

their fellow-miners as well as their own; and, in case of accident through any neglect on their part, the law will hold them responsible. The first preventive means which should be insisted upon is a really efficient and "safe safety-lamp." Many of the so-called safety-lamps are no safety-lamps at all: in fact, they are apt "to lull the inexperienced into false security." Hundreds of miners imagine, that, if they have a safety-lamp, they can work with impunity in the most fiery veins, and can defy almost any risk. "We all know how dangerous such false security must be, and how treacherous many of the lamps have proven when a sudden 'blower' of gas has been struck. Again: many of the lamps can be picked by an ingenious collier, and the most disastrous consequences have resulted from this tampering with so-called safety-lamps. Whatever may be said to the contrary, the mining-world still wants a really safe safety-lamp, — one which, while giving a good light, will defy all tampering by the collier, and resist any amount of explosive gas with which it may come in contact. Until we have this, we may look in vain for any appreciable decrease in the number of explosions in our fiery pits, and the lives of our colliers will be more or less in jeopardy."

Another provision, which should be enforced under

every circumstance, is to prohibit the use of gunpowder in all fiery, bituminous mines. Shot-firing has, in all probability, to answer for more fatal casualties than some are inclined to ascribe to it. This does not happen so much in anthracite mines, from the fact that there is scarcely any dust floating in the air when compared with bituminous mines: however, we are not sure but that it may apply in some instances even to them.

It is certain that people are maimed and burned by blasting, at distances varying from ten to a hundred and eighty yards, when there is no fire-damp present to cause such destruction; then, it is quite clear that this results, either from the simple force and flame of the shot on account of the weight of the charge, or from this force and flame assisted by the rapid combustion of coal-dust as it travels on its course, or from the force and flame assisted by an instantaneous emission of gas, in consequence of a partial vacuum being formed by the rushing blast. With a view of testing these assumptions, careful experiments were made, a description of which may be found in "Colliery Guardian," England, p. 13, vol. xxxii., the summing-up of which is as follows: —

"1. The flame from a blown-out shot, unassisted by gas or coal-dust, does not travel farther than five, or, at the utmost, ten yards, entailing little or no danger.

"2. If coal-dust be present, even in a comparatively damp mine, the flame may not travel fifty yards. That in a dry mine of a high temperature this distance would be greatly exceeded; and since miners, as a rule, consider themselves safe at from fifteen to twenty yards from the point where the powder is used, a blown-out shot under these circumstances is a source of great danger.

"3. That the violence of the blast from either gunpowder or fire-damp is much increased when coal-dust is present.

"4. That, on any partial vacuum being formed in an underground coal-working, fire-damp will instantly issue in dangerous quantity; and there are fair grounds for assuming that a shot blowing out in the face of a narrow heading, and setting coal-dust on fire in its course, would, by its exhaustive action, produce such a vacuum, and might cause a serious explosion in a mine practically clear of gas.

"5. Although no experiments have been made directly to test the result of coal-dust set on fire in air heavily loaded with fire-damp, there is every likelihood that such an occurrence would be attended with grave consequences.

"6. That it is desirable that any system of blasting coal which entails heavy charges of gunpowder, and an unusual liability to 'shots blowing out,' such as blasting without side-cutting, or nicking, or using improper materials for stemming, should be discontinued.

"7. A large body of flame, such as results from a very heavy charge or from a blown-out shot, is required to ignite coal-dust; that in blasting with charges not exceeding twelve ounces, accompanied by the proper preparation of holing and side-cutting, there is little liability of this taking place.

"To discard all shot-firing means, in many mines, a considerable increased cost in the working of coal. But life is the first consideration, and the safety of the collier should be the one great object of the proprietor and manager. The opinion of the best mining engineers is, that so long as shot-firing is allowed, even under the most favorable circumstances, so long will there be a certain amount of risk, while, in many cases where the plan is adopted, it often leads to most serious and fatal consequences."

36. The third point, which should be earnestly considered, is efficient ventilation. There is no country in the world where such facilities are offered for good ventilation, because of the thick coal-seams; yet in very many instances managers appear to rely too much upon this advantage, and fail to conduct the air properly in its journey through the mines; and there are some that could not, although by law required, give the inspector a design of their ventilation. Thus it is not at all to be wondered at, when such gross negligence prevails, that accidents now and then occur. The cheapest method of obtaining more air is by splitting, and mechanical means.

Enlarging the airways lessens the velocity, and is another, though costlier, mode of obtaining more air: in small seams it is, however, absolutely imperative.

CHAPTER X.

MEASURING THE AIR.

37. To measure air travelling through mines, various methods have been employed: those in most general use may be classed under three heads; viz., —

(*a*) Travelling at the same velocity as the air-current, and noting the distance passed over in a unit of time.

(*b*) Determining from observation the rate at which small floating particles are carried along by the current, and assuming their velocities to be identical with that of the air-current itself.

(*c*) By the use of the anemometer, or other instruments.

(*a*) By the first method it is necessary to select as regular an airway as possible, in which is measured a length of from fifty to two hundred yards. This is traversed in the same direction as the current, a candle being held in the hand, whose flame must be kept vertical. The time of walking the distance, which is the same as the velocity of the air, is measured by a watch. The mean of two or three experiments gives a rough estimate of the velocity.

(*b*) By light bodies, such as down or powder-smoke, and noting the time it takes for the particles to pass from one point of the gallery to another. Under this head may be classed measurements made by ammonia, sulphuric-ether, etc. To measure air by volatile liquids, small phials containing the liquid are broken at a certain place in the airway, and the time noted that it takes to pass from this point of the gallery to another, the distance between the two having been previously ascertained.

By vapor-measurements Mr. Arnold constantly obtained the same results. This method, so simple in practice, is more exact than measuring by powder-smoke, down, or other light substances. Mr. Arnold considers it to be less subject to error than the best anemometer yet invented.

On the other hand, others claim that — the sense of

smell being more acute in one person than in another, and different for the same persons at different times — it is impossible to measure the air as accurately as with an instrument.

Many experiments made with anemometers show a variation in their co-efficient at different velocities of the air-current. M. Guibal, in order to suppress this source of error, sought an approximate velocity; then, by using the co-efficient thus found, he made new experiments, which would give the true co-efficients. Two anemometers are required to do this: if they agree, the measurement is correct; if they do not, one of the two is wrong, and it is then necessary to ascertain which.

(c) Miners have long recognized the importance of knowing the volume of air for all pressures of the atmosphere: hence they regularly measure the air traversing all the main intakes in the returns and at certain points in the various districts. Some use powder-smoke; but that test has been practically superseded by the anemometer: however, whatever method is employed, the measurements are, or should be, made at certain prescribed times. There are many and various instruments for measuring the velocity of the air, among which may be mentioned Devillez's, Dickinson's, Briam's, Robinson's, Casella's, and Casartelli's anemometers. Briam's is the one most generally used in this country, and is a modification of Robinson's as originally made for meteorological use. It consists of a series of vanes, which revolve by the action of wind. Each revolution is transmitted to dials by means of wheels and pinions. These instruments are made of various sizes, from four to twelve inches. The dials are six in number, marked for feet, hundredths, thousandths, etc. Whatever instrument is used, all that is required to ascertain the velocity is to read the figures on the respective dials before and after experiment, then to subtract the first from the second; to the remainder is added a value corresponding to the constant friction, and which will be found with the table that comes with each instrument.

The special formula is of the form

$$V = ar + u$$

in which r equals the number of revolutions per minute, a equals constant proportional to the number of linear feet traversed by the air in a revolution, and u represents the losses due to friction.

For the anemometer generally used in Pennsylvania, the correction is about thirty feet when the instrument is new and clean; but the dirt and grit, to which it is more or less exposed, have, no doubt, a tendency to increase the friction, or the correction in time. To determine the amount of correction required, the instrument is placed on a whirling table; the anemometer is whirled around by the table revolving; the velocity of the table is then compared with the indications of the anemometer.

38. The method of procedure in conducting experiments to find the useful effect of fan-ventilation varies materially from the ordinary method of ascertaining the velocity of the air, because the revolutions of the fan, the indicated power of the engine, water-gauge, etc., must be considered.

The air is measured by the anemometer in the ventilator-drift if possible. At the place of measurement, strings or wires should be fixed so as to divide the drift into, say, ten divisions of nearly equal area. The anemometer should run at least one minute in each division; one minute interval should be taken for reading the instrument, and moving it to the next division. Simultaneously with the air-measurement, diagrams should be taken from the engine at intervals of three or five minutes. Each diagram should be accompanied with an observation of the water-gauge, and the number of strokes per minute of the engine-piston. The usual working-speed should be adopted for the experiment, and it should be maintained as uniformly as possible throughout the trial. After completing the drift-measurements, a second air-measurement should be made, either in the intake or return airways, to check, in some degree, the drift-measurement. To make these check-measurements, move the anemometer uniformly over the whole area of the airway, for, say, two minutes, repeating the observation twice to avoid error. During these check-measurements, diagrams of engine, and observations of speed and water-gauge, should be taken, as in the first measurement.

The laws of Mariotte and Gay-Lussac may be applied to correct the volume of air measured in the intake or return airways to the condition of the ventilator-drift at the surface; namely, for pressure and temperature, as follows: —

Problem. — Supposing the volume of air measured at the intake to be 100,000 cubic feet per minute, the required volume which it would occupy in the ventilator-drift is found to be 107,900 cubic feet with the following conditions: —

	Barometer.	Temperature.
Ventilator-drift	30.30″	60° F.
Intake airway	31.25″	37° F.

Neglecting any small increase of volume due to

evolution of gas, or absorption of aqueous vapor in the mine, we have in the ventilator drift,

$$100000 \times \frac{31.25'' \times (60° + 461)}{30.30'' \times (37° + 461)} = 107900 \text{ cubic feet.}$$

In order to find the amount of work which is expended in producing ventilation, and what amount is lost in overcoming friction, it is necessary to use an indicator, the diagram of which will give us the effective horse-power, which differs from the nominal or theoretical horse-power.

Problem. — Find the per cent of power used by a fan with the following data: —

Area of piston 484 inches; stroke 2 feet; speed 60 revolutions, or 240 feet of piston per minute; indicated effective pressure on piston 20.89 pounds per square inch; then,

$$\frac{484 \times 240 \times 20.89}{33000} = 73.54 \text{ horse-power engine.}$$

The air in fan-drift measures 106,680 cubic feet per minute. Water-gauge in fan-drift measures 2.8″. Then

$$\frac{106680 \times 2.80 \times 5.2}{33000} = 47.06 \text{ horse-power in the air.}$$

Therefore the per cent of power utilized will be

73.54 : 47.06 :: 100 : 64 per cent. *Ans.*

THE BAROMETER.

39. It is supposed that Torricelli derived from Galileo the definite conception of atmospheric pressure. Pascal, however, was first to state that the mercurial column decreased in length as we ascend. This experiment was for the first time performed at Clermont, on the top of the Puy de Dôme, Sept. 19, 1648. The barometer in its simplest form consists of a tube, about thirty-four inches long, closed at the top. This tube is filled with mercury, then inverted in a vessel containing mercury. The atmospheric pressure on the vessel of mercury will force the mercury up the tube, or let it sink, according as that pressure is greater or less. These risings and fallings are measured by means of a scale. As mercury expands by heating, it follows that a column of warm mercury exerts less pressure than a column of the same height at a lower temperature. It is usual, on this account, to reduce the actual height of the column to the height of a column of mercury at the temperature of freezing water, which would exert the same pressure.

The formula for this correction is

$$h_0 = h - hm(t - 32°),$$

in which $h =$ height of mercury at $t°$, $h_0 =$ height of

mercury at freezing-point, m = co-efficient of expansion of mercury per degree Fahrenheit = $\frac{1}{9990}$ = 0.0001001.

When very exact readings are required, corrections must be made for expansion of the scale by which the height of the mercurial column is measured, also for capillarity.

(a) The Aneroid Barometer is a thin metallic vessel partially exhausted of air, and sealed: consequently it will expand or diminish in size as the atmosphere is lighter or heavier. This change in size, M. Vidi made use of, and transmitted the movement to an index. The Aneroid Barometer is a very convenient instrument; as it is round, and of small compass. In the second geological survey of Pennsylvania it was used, to a great extent, for determining heights, and making contours, in the anthracite coal-regions. Good Aneroid Barometers are compensated for differences of temperature.

(b) Atmospheric pressure will, according to the condition of the weather, vary from 28.5 to 31 inches of mercurial column. When the barometer rises, the thermometer usually falls, and *vice versa*. The discharge of gas becomes greater when the barometer falls, because the atmospheric pressure which before kept the gas pent up is lessened; and hence, wherever the pressure of gas is strong enough to overcome the lessened atmospheric pressure, it escapes. The barometer is useful, therefore, as it will give warning when an increased discharge of gas will take place; and hence precautions may be taken to overcome it by increasing the volume of air.

Sudden falling of the barometer is much more dangerous than a gradual fall; for in the first case more gas will be given off in less time than in the second. When the barometer is 27 inches, the pressure of the atmosphere per square foot is 1,908.23 pounds; at 28 inches, it is 1,978.90 pounds; at 29 inches, it is 2,049.58 pounds; at 30 inches, it is 2,120.25 pounds; at 31 inches, it is 2,190.93 pounds.

The following table will be found useful in order to ascertain the pressure per square foot, or fractional part of a foot, for a given height of the barometer.

TABLE VII.
PRESSURE OF AIR PER SQUARE FOOT.

Inches.	Pounds.	Inches.	Pounds.	Inches.	Pounds.
0.01	0.71	0.12	8.48	0.50	35.34
0.02	1.41	0.13	9.19	0.60	42.41
0.03	2.12	0.14	9.90	0.70	49.48
0.04	2.83	0.15	10.60	0.80	56.54
0.05	3.53	0.16	11.31	0.90	63.61
0.06	4.24	0.17	12.02	1.00	70.68
0.07	4.95	0.18	12.72	2.00	141.36
0.08	5.65	0.19	13.43	3.00	212.04
0.09	6.36	0.20	14.14	4.00	282.72
0.10	7.07	0.30	21.20		
0.11	7.77	0.40	28.27		

Problem. — Require the amount, in cubic feet, of air and gas that may be expected to be given off for 1,000 cubic feet of open space in the goaves, or other waste places, by a falling of the barometer from 30.4 inches to 28.75 inches.

The pressure at 30.4 inches = 2,148.52 pounds
The pressure at 28.75 inches = 2,031.91 pounds

Difference . . . . . . 116.61 pounds.

Then,
$$2{,}148.52 : 116.61 :: 1000 : 54.27,$$
or
$$30.40 : (30.40 - 28.75) :: 1000 : 54.27$$

cubic feet of gas, which, theoretically, may be given off by a reduction of pressure equal to that indicated above.

Experience in the use of the barometer in mines has shown that its indications are from one to three hours behind what is actually taking place. On this account it has been asserted that the barometer is not to be relied upon to give warnings. According to the "Colliery Guardian," Jan. 31, 1883, the government issued and sent out in England, during the year 1882, thirty-two warnings, nineteen of which were justified by subsequent events. Twelve were followed in three days by explosions which caused one hundred and thirty-nine deaths; two were followed on the fifth day by explosions causing forty-three deaths; twenty-three lives were lost on the sixth day after the warning — showing that a total of two hundred and five lives were lost in six days from the issue of the warnings, while five lives were lost on the day of the issue.

We have noticed, when there have been explosions telegraphed from England, that, within a short time, explosions have occurred in our American mines. This warning should never be slighted, whether the explosions are due to falling or rising barometer. These teachings, we are aware, conflict with some of the late writings on this point; but we are so convinced by analogy that one of our very best warnings is the barometer, that we hope our own government will follow England's Signal Service Bureau, and send our colliery managers warnings.

CHAPTER XI.

CONCLUSION.

40. THE first method employed to ventilate mines was, we believe, to agitate the air by shaking a cloth, next, natural ventilation by means of upcast and downcast shafts. This method was followed from necessity by the furnace; and even to-day this latter method is considered by some the best, on account of the liability of mechanical ventilators to get out of order, and so stop the current. The furnace is still used in many mines, but is being gradually superseded by the mechanical means now at our command. It is, however, more effective in deep mines than any thing as yet employed, but it is just in this position that it is the most dangerous, as, the deeper we descend, the more gas we are likely to encounter. In shallow mines, especially if worked at a dip, the fan is the more economical of the two systems. The steam-jet, at one time, was

a rival of the furnace. This was followed by allowing water to fall down the downcast; but this system proved inefficacious, as it did not provide enough change of temperature, and the water had to be pumped back again, in most cases, out of the mine. The mechanical ventilators of late years have been numerous; but, of all the number, not one perfect one has been produced. Guibal's Fan seems to be the favorite; and to him is due the simplest, and, with his sliding shutter, the most effective fan. Nixon made a ventilator on the air-pump system, the immense piston of which goes backwards and forwards on wheels. The air is received into a chamber, and forced out by this piston: it works exactly upon the same principle as a pump, drawing the air from the mine by one set of valves, and then forcing it out by another on the back movement of the piston.

Mr. Struve constructed a machine upon a principle similar to that of Mr. Nixon, using two large gas-tanks, arranged with valves. These gasometers moved up and down alternately in water, this means taking the place of a piston. There are also centrifugal machines, receiving the air at the centre, and throwing it off from the ends of the blades: others are made on the windmill plan, each blade, as it revolves, cutting out a definite portion of the air. The Champion Ventilator, the only really distinctive American ventilator, is so arranged as to be used either as a forcing or exhausting fan. It is credited with very good results. Every new fan which has been built of late has been declared to give at least ten per cent more air for the same amount of power than any fan previously invented. This can hardly be believed, unless it has been proven by placing the new fan in a position where some other fan has been, with the conditions of the mine and the power just the same.

Theoretical comparison of two fans at different mines cannot give any thing like exact results, as the airways and resistances will not be similar in the two mines; and, while one fan may be better than the other, yet, in its position, it may be unable to cope with an inferior adversary more favorably situated. A larger water-gauge may be obtained in the fan-house than in the airway leading to the fan: this is accounted for by the fact that the fan offers more or less resistance to the air, and slightly impedes its discharge. A perfect fan should not do this, — at least to any great extent. Theoretically and practically the amount of ventilation obtainable from furnace-action will depend upon the difference in weight of two air-columns. An improper consideration of this subject has led the enemies of the furnace to state that there is a material difference between the action of furnace and fan ventilation; the former being likened to propulsion, the latter to traction. Were the air propelled, the power expended would be applied to force the air down one of the shafts, which the furnace does not do, but draws the air down, expelling it in lighter form, the same as the exhaust fan.

41. Suppose the exhaustion produced in a fan-drift at 40 revolutions per minute of the fan averages about 1.25 inches of water-gauge, while at 60 revolutions of the fan we have 2.8 inches water-gauge, — a rise proportional to the square of the speed. Since the vacuum increases as the square of the number of revolutions per minute, the quantity of air produced should be the same per revolution at any speed where the conditions are unchanged, for the volume of air varies as the square root of the water-gauge; i.e., the square root of the lowest pressure bears the same relation to the square root of the highest pressure as forty revolutions bear to sixty revolutions per minute; or

$$\sqrt{1.25} : \sqrt{2.8} :: 40 : 60.$$

Therefore, if it were desirable to pass double the quantity of air though a mine or drift where the existing friction is equal to one inch of water-gauge, without making any alteration in the underground arrangements, the effect of the change would be to increase the measure of resistance to four inches of water-gauge; also the power required to overcome this friction would be eight times that employed for the original quantity, as, in addition to the friction being fourfold, the volume of air is also doubled. In like manner, for three times the quantity, we have nine times the resistance, and require twenty seven times the power.

The following table is taken from Mr. R. Howe's paper, printed in the Transactions of Chesterfield and Derbyshire Institute of Engineers; the fan under consideration being a Guibal. In the preparation of the table, the following general principles are observed: —

1st, The quantity of air increases in proportion to the speed of the fan.

2d, The water-gauge increases proportionately to the square of the number of revolutions of the fan.

3d, The horse-power in the air increases as the cube of the quantity.

4th, The steam pressure in the cylinder is in proportion to the square of the piston's speed.

5th, The horse-power of the engine is proportionate to the cube of the number of revolutions per minute, or to the cube of the volume of air.

Illustrations: —

1st, 25 rev. : 30 rev. : : 44450 : 53340 cubic feet of air.
2d, $(25)^2$ rev. : $(30)^2$ rev. : : 0.48 w.g. : 0.69+ or 0.7 w.g.
3d, $\dfrac{44450 \times 0.48 \times 5.2}{33000} = 3\ 4$ horse-power in the air.
4th, $(25)^2 : (30)^2 : : 3.62 : 5.21+$ pressure on piston.
5th, $(25)^3 : (30)^3 : : 5.32 : 9.19$ effective HP. of engine.

Assuming the quantity of air discharged for each revolution of the fan to be 1,778 cubic feet, then, at 25 revolutions per minute, the number of cubic feet passing will be

$$25 \times 1778 = 44450 \text{ cubic feet.}$$

This does not always hold good when the number of revolutions are greatly increased, from the fact that the baffling of the air does not admit the water-gauge to register correctly. "The theoretical quantity of a fan at High Colliery was short, when compared with the measurements, 3,048 cubic feet per minute." The fan gave off 1,584 cubic feet per revolution: therefore the quantity that should have been delivered was 82,368; but from measurement 81,923 cubic feet were delivered at 1.3 inches water-gauge. From the formula

$$Q = \sqrt{\dfrac{h}{h'}} \times R$$

it was found that while

$$\sqrt{\dfrac{1.9}{1.3}} \times 82.36 = 99.665 \text{ cubic feet}$$

§ 42. MINE VENTILATION. 115

was all that the water-gauge called for, the measured quantity was 102.713 cubic feet, making a difference in the second case of 3.048 cubic feet per minute, as stated above.

TABLE VIII.
GUIBAL FAN.

Number of revolutions per minute.	Inches water-gauge in fan-drift.	Quantity = $R \times 1{,}778$ discharged per minute.	Horse-power of air in fan-drift.	Effective pressure of steam in pounds per square inch on piston.	Effective horse-power of engine.
25	0.48	44,450	3.40	3.02	5.32
30	0.70	53,340	5.88	5.22	9.10
35	0.95	62,230	9.29	7.11	14.50
40	1.24	71,120	13.04	9.28	21.78
45	1.57	80,110	19.85	11.75	31.02
50	1.94	88,900	27.23	14.50	42.55
55	2.35	97,790	36.25	17.55	56.04
60	2.80	106,680	47.00	20.89	73.54
65	3.28	115,570	59.84	24.51	93.50
70	3.81	124,460	74.34	28.43	116.77
75	4.37	133,350	91.93	32.04	143.63
80	4.98	142,240	111.57	37.13	174.31
85	5.62	151,130	133.82	41.93	209.08
90	6.30	160,020	158.85	47.00	248.20
95	7.02	168,910	186.83	52.37	291.90
100	7.77	177,800	217.90	58.03	340.46

42. The Guibal Fan belongs to that class of ventilators called centrifugal, because of the air being thrown

off at the tip of the blade tangent to the circumference of the fan. There are probably more of these fans in use than of any other kind: seemingly they have the precedence among the best mining engineers, maybe from their simplicity of construction, or from their non-liability to get out of order. The committee of the North of England Institute, appointed to determine the useful effect of different fans, reported: —

	Per cent useful effect.
Struve	57.80
Guibal	52.95
Waddle	52.79
Schiele	49.27

Thus placing them nearly on a par with a displacement machine in the van. Mr. Howe, in his experiments quoted above, places the useful effect of the Guibal at about sixty-four per cent. The fan is enclosed in a house, the air being discharged through a chimney which gradually expands towards the top. That there is a certain amount of benefit derived from this chimney has been proved by experiment, simply because by its use there is a saving in final velocity by the opportunity it affords the air for expanding. The opening for the discharge of the air is regulated by an adjustable shutter. To find the most advantageous position for this shutter, a series of experiments must be instituted.

First, the shutter is fixed at a certain point, then the amount of air ascertained by measurement, as before explained; the shutter is now lowered and the measurement again taken; again the shutter is lowered and the measurement taken. If, however, in the last position, a less amount of useful effect is shown than in the previous position, the shutter has been lowered beyond its most useful position, and must be returned to the second position.

The position of the shutter depends upon the speed at which the fan runs. Should the shutter vibrate when the speed is increased, it shows that the fan is not working properly, and that the shutter needs regulating.

A modification of the Guibal Fan is used in the anthracite regions, having, instead of the shutter, a spiral casing, commencing at the orifice of discharge (throat), and extending sometimes the whole circumference, according to the notion of the engineer. Whenever this spiral casing is too short, thumping will take place. The discharge is regulated at the throat by a short shutter, or by nailing boards over the orifice. The more recently built fans are made as large as forty-five diameter, and have fire-proof casings of iron.

COMPARATIVE ECONOMY OF FURNACE AND FAN VENTILATION.

(FROM MR. R. HOWE'S PAPER.)

43. Suppose we have two furnace-pits, the first 260 feet deep, the second 655 feet deep.

To arrive at the horse-power of furnace-ventilation, we must find the pressure producing it.

First, Barometer of the first pit, 30.3″; temperature of downcast, 55°; of upcast, 240°. The pressure, therefore, of the downcast in pounds per square foot, is

$$\frac{1.3253 \times 30.3}{459 + 55} \times 260 = 20.312 \text{ pounds.}$$

The pressure of the upcast is

$$\frac{1.3253 \times 30.3}{459 + 240} \times 260 = 14.936 \text{ pounds.}$$

$20.312 - 14.936 = 5.376$ pounds as the pressure per square foot for ventilating pressure, and this pressure will give us, say, 30,358 cubic feet of air, then the horse-power will be

$$\frac{30358 \times 5.376}{33000} = 4.94 \text{ horse-power in the air.}$$

Second, Mean barometer, 30.6″; temperature of downcast, 58°; temperature of upcast, 117°; then the pressure in pounds per square foot is, for downcast,

$$\frac{1.3253 \times 30.6}{459 + 58} \times 655 = 51.379 \text{ pounds};$$

for upcast,

$$\frac{1.3253 \times 30.6}{459 + 117} \times 655 = 46.116 \text{ pounds}.$$

Hence, for ventilating pressure, $51.379 - 46.116 = 5.263$ pounds per square foot; and this pressure gives 48,230 cubic feet of air per minute, which, reduced, gives us 7.69 as the horse-power in the air.

These two furnaces ventilated the Hollingwood pits. The No. 1 furnace used 3 tons 12 hundred weight of coal in 24 hours, or 68 pounds per hour per horse-power in the air. No. 2 furnace used 3 tons 1 hundred weight per day, or 37 pounds of coal per hour per horse-power in the air. The two furnaces circulated 78,588 cubic feet of air, with a horse-power of 12.63.

A Guibal Fan was afterwards substituted to take the place of these furnaces, the ordinary speed of which was sixty revolutions per minute. The average quantity of air circulated at this speed was 106,680 cubic feet; the pressure, 2.8 water-gauge, or 14.56 pounds per square foot.

Tabulated results of the above calculations show the economy of fan over furnace ventilation. The wages and price of coal are those regulated by the English market at the time of the writing, reduced to United-States money.

TABLE IX.

	Quantity by the two furnaces.	Quantity at 60 revolutions of fan per minute.	Due to Fan.	
			Increase.	Decrease.
1. Cubic feet of air per minute .	78,588	106,680	28,092	--
2. Pressure in pounds per square foot	5.304	14.56	9.256	--
3. Horse-power in air, including shaft friction . .	12.63	47.06	34.43	--
4. Cost of fuel, wages, etc., for 24 hours	$10.44	$5.33	--	$5.11
5. Cost for 24 hours per horse-power in the air	$0.82	$0.12	--	$0.70
6. Pounds of coal consumed per horse-power in the air per hour	49	10	--	39
Amount saved in one year by fan $2,120.65				

APPENDICES.

APPENDIX A.

FORMULAS.*

44. Let a = area of airway in square feet.
o = perimeter of airway in feet.
l = length of airway.
s = rubbing-surface in feet.
k = co-efficient of friction, 0.0217 of a pound at a velocity of 1,000 feet per minute.
p = pressure in pounds per square foot of sectional area.
v = velocity of the air in feet per minute.
w = water-gauge.
q = quantity of air circulating, in cubic feet, per minute.
u = units of work applied to circulate the air.
HP = horse-power of ventilation.

---

* From Mining Herald Almanac.

MINE VENTILATION. § 44.

Then the following formulas for friction of air in mines may be deduced: —

1. $a = \dfrac{ksv^2}{p} = \dfrac{pa}{p} = \dfrac{ksv^2 q}{u} = \dfrac{ksv^3}{pv} = \dfrac{u}{pv} = \dfrac{q}{v} = \dfrac{q}{\sqrt[3]{\dfrac{u}{ks}}}.$

2. $o = \dfrac{s}{l}.$

3. $l = \dfrac{s}{o}.$

4. $s = lo = \dfrac{pa}{kv^2} = \dfrac{u}{kv^3} = \dfrac{qp}{kv^3} = \dfrac{vpa}{kv^3}.$

5. $k = \dfrac{pa}{sv^2} = \dfrac{u}{sv^3} = \dfrac{p}{\dfrac{sv^2}{a}} = \dfrac{w5.2}{\dfrac{sv^2}{a}}.$

6. $p = \dfrac{ksv^2}{a} = \dfrac{u}{q} = \dfrac{pa}{a} = \dfrac{ksv^3}{q} = \dfrac{u}{av} = 5.2w.$

7. $v = \dfrac{u}{pa} = \dfrac{q}{a} = \sqrt[3]{\dfrac{u}{ks}} = \sqrt[3]{\dfrac{qp}{ks}} = \sqrt[2]{\dfrac{pa}{ks}}.$

8. $w = \dfrac{\dfrac{ksv^2}{a}}{5.2} = \dfrac{p}{5.2}.$

9. $q = va = \dfrac{u}{p} = \dfrac{ksv^3}{p} = \dfrac{\sqrt{pa}}{ks} \times a = \sqrt[3]{\dfrac{u}{ks}} \times a.$

10. $u = qp = vpa = \dfrac{ksv^2 q}{a} = ksv^3 = q\,5.2\,w = HP \times 33000.$

§ 44. MINE VENTILATION.

11. $HP = \dfrac{u}{33000} = \dfrac{qp}{33000}.$

12. $v^2 = \dfrac{pa}{ks} = \left(\sqrt[3]{\dfrac{u}{ks}}\right)^2.$

13. $v^3 = \dfrac{u}{ks} = \dfrac{qp}{ks} = \dfrac{vpa}{ks}.$

14. $pa = ksv^2 = \left(\sqrt[3]{\dfrac{u}{ks}}\right)^2 \times ks = \dfrac{u}{v}.$

"These formulas comprise the pressure due to resistance, and not that necessary for final velocity: they are, therefore, more correct for long than for short airways. The pressure required for final velocity becomes a smaller fraction of the whole drag as the airways extend. If it be required to take into account the pressure to create velocity,

Instead of using $a = \dfrac{ksv^2}{p}$, use $\dfrac{ksv^2}{p - P}.$

Instead of using $pa = ksv^2$, use $a(p - P) = ksv^2.$

Instead of using $p = \dfrac{ksv^2}{a}$, use $p - P = \dfrac{ksv^2}{a}$, or $\dfrac{ksv^2}{a} + P.$

Instead of using $s = \dfrac{pa}{kv^2}$, use $\dfrac{(p - P)a}{kv^2}.$

Instead of using $v = \sqrt{\dfrac{pa}{ks}}$, use $\sqrt{\dfrac{(p - P)a}{ks}}.$

APPENDIX B.

PROBLEMS.

45. 1. What is the area of an airway 6 feet by 5 feet? What is its perimeter? $a = 30$; $o = 22$. *Ans.*

2. What is the area of a shaft 14 feet diameter?
Rule. — Diameter^2 × 0.7854.   153.98 + feet. *Ans.*

3. What is the perimeter of a shaft 16.5 feet diameter?
Rule. — Diameter × 3.1416.   58.83 + feet. *Ans.*

4. The long axis of an elliptical shaft is 14 feet, its short axis 6 feet: what is its area?
Rule. — $A \times a \times 0.7854$.   65.9736 feet. *Ans.*

5. Find the perimeter of an elliptical shaft whose axis A is 16 feet, and its axis a 8 feet.

$$Rule. - \frac{\sqrt{\frac{A^2+a^2}{2}} + \frac{A+a}{2}}{2} \times 3.1416.$$

38.7 feet. *Ans.*

6. An air-course is 500 yards long, 6 feet high, and 7 feet wide: what is its area, perimeter, and rubbing-surface?
$a = 42$; $o = 26$; $s = 39000$ square feet. *Ans.*

7. What is the rubbing-surface of a shaft 15 feet diameter, 1,200 feet deep?   56548.8 feet. *Ans.*

§ 45. MINE VENTILATION.

8. In an airway 8 feet by 9 feet, when the current has a velocity of 15 feet per second, what quantity of air is passing per minute?

Rule. — $a \times v \times 60''$.   64800 cubic feet. *Ans.*

9. When the water-gauge is 1.85 of an inch, what pressure per square foot does it indicate?

Rule. — $w \times 5.2$.   9.62 pounds. *Ans.*

10. When the quantity of air passing is 60,000 cubic feet, with a water-gauge of 1.5 inches, what are the units of work producing ventilation?

Rule. — $u = pq$.   468000 units. *Ans.*

11. What horse-power is there in 468,000 units of work?

Rule. — $HP = \dfrac{u}{33000}$.   14.18. *Ans.*

12. The pressure producing ventilation is 7.8 pounds: what is the water-gauge?

Rule. — $w = \dfrac{p}{5.2}$.   1.5 inch. *Ans.*

13. There are 50,000 cubic feet of air passing, having a rubbing-surface 24,000 feet, and an area of 20 square feet: what is the water-gauge?

Rule. — $w = \dfrac{\frac{ksv^2}{a}}{5.2}$.   3.12 inches water-gauge. *Ans.*

14. The rubbing-surface of an airway is 25,000 feet, its area 25 square feet: what is its length?

1250 feet. *Ans.*

15. What units of work are necessary to overcome friction of an airway 6 feet by 6 feet, 1,000 feet long, when the quantity passing is 7,200 feet per minute?

Rule. — $u = \dfrac{ksv^2}{a} \times q.$ 4166 units. *Ans.*

16. Let $a = 36$; $s = 24,000$, to find the value of
$$\dfrac{1}{\sqrt{\dfrac{(\frac{1}{a})^2 \times s}{a}}}.$$
1.39. *Ans.*

17. With 0.9 of an inch water-gauge, 16,000 cubic feet of air are passing: what quantity will pass when there is a water-gauge of 1.6 inches?

$\sqrt{0.9} : \sqrt{1.6} :: 16000 : 21333 +$. *Ans.*

18. With a fan and furnace combined, 46,706 cubic feet are produced; the furnace produces alone 42,670 cubic feet: what will the fan produce by itself?

$\sqrt{46706^2 - 42670^2} = 18993.$ *Ans.*

19. If, with a water-gauge of 0.65 of an inch, 20,000 cubic feet of air are obtained, what height will the water-gauge be when there is a quantity of 75,000 cubic feet of air passing?

$\dfrac{q}{q_1} = x \therefore x^2 \times w = 9.139.$ *Ans.*

20. How much must we increase the pressure to double the quantity of air?   4 times. *Ans.*

21. How much has the ventilating power to be increased to treble the quantity of air?
$$3^3 = 27 \text{ times. } Ans.$$

22. If we obtain 25,000 cubic feet of air by 5-horse power, what horse-power will be required to circulate 60,000 cubic feet in the same mine?
$$(q')^3 : (q)^3 :: 5HP : Ans.$$
$$\text{or } q' : q :: \sqrt[3]{5} : \sqrt[3]{Ans.}$$

23. There are two air-courses through which a total quantity of 100,000 cubic feet of air is passing; the resistances of the airways are in the proportion of 4 to 1: what quantity will pass along each?

$$\frac{\sqrt{1} \times 100000}{\sqrt{4} + \sqrt{1}} = \frac{100000}{3} = 33333\tfrac{1}{3} \text{ cubic feet for airway}$$
having greater resistance.

$$\frac{\sqrt{4} \times 100000}{\sqrt{4} + \sqrt{1}} = 66666\tfrac{2}{3} \text{ cubic feet for airway having}$$
only one-fourth the resistance.

24. Find the motive-column where the upcast and downcast shafts are 540 feet deep, the temperature of the upcast being 129°, and that of the downcast 43°.

Rule. — $M = D \times \dfrac{129 - 43}{459 + 43}.$ 92.51. *Ans.*

APPENDIX C.

46. The quantity of air required per man for respiration has been variously estimated by the following authorities: —

" The volume of air contained in the lungs, accordingly is 109 cubic inches; after respiration, 60 cubic inches remain in the chest; total volume 170 cubic inches. Amount of each inspiration has been differently estimated, it is probably 16 to 20 cubic inches." — Mr. Goodwin.

" Men between five and six feet in height, after a complete inspiration, expel by force, on an average, 225 cubic inches at 60°. This is called 'vital capacity of the lungs.' " — Mr. Hutchinson.

" Assuming a man takes twenty breaths per minute, each 40 cubic inches vitiates 28 cubic feet per hour. Besides this, a quantity of vapor is emitted, which, according to Dumas, amounts to 0.0836 of a pound of water per hour, — enough to saturate 7.1 pounds of air at 60°. And if we allow, that, to be healthful and pleasant, the air should be only one-half saturated, we require 14.2 pounds of air, or 187 cubic feet, giving us a total of 215 cubic feet per hour, which happens to be the capacity of a 6-foot cube. This is the maximum quantity necessary for clean, healthy persons. For prisons, etc., it should not be less than 350 cubic feet, and for hospitals 1,000 cubic feet, per hour per head." — Mr. Box.

"A minimum of 100 cubic feet per minute for each man and boy, for sanitary purposes alone." — Mr. Herbert Mackworth.

"From 100 to 500 cubic feet per minute for each collier, according to condition of the mine." — Mr. Hedley.

"The minimum quantity of fresh air for the most harmless of pits ought to be from 10,000 to 15,000 cubic feet per minute." — Mr. Dunn.

"In most fiery mines, an average of 600 cubic feet per minute per collier is circulated, and nearly 200 cubic feet per minute for each acre of waste." — Professor Phillips.

"For all anthracite mines, nearly double the above estimates (which are for bituminous mines) should be allowed, because of the much greater volume of powder-smoke, due to the large amount of blasting done." — Thomas J. Foster.

The Mine Ventilation Act for the anthracite region of Pennsylvania provides for 66 cubic feet of pure air per minute for each man working. All authorities agree in declaring this amount inadequate.

MINE VENTILATION.

Authority.	Lung capacity.	No. of respirations per minute.	Cubic feet inhaled per hour.	Volume carbonic acid gas respired per hour.	Cubic Inches inhaled per minute.
Roscoe { youth	21	15	11	0.33	315
Roscoe { adult.	43	15	22	0.66	645
Atkinson	–	–	12	0.84	345
Colliery Pocket-Book	–	–	16	0.54	480
Briston M. S. Lectures	–	–	28	0.84	–
Annales des Mines	–	–	33	1.00	–
Average	–	–	22	0.72	–

APPENDIX D.

ASPHYXIA.

47. THE miners are exposed to asphyxia when the circulation of the air is not sufficiently active, and when they imprudently penetrate into ancient and abandoned workings, or wherever the air has not sufficient oxygen.

The symptoms of asphyxia are sudden cessation of the respiration, of the pulsations of the heart, and of the action of the senses. The face is swollen and flushed;

the eyes protrude; the features are distorted, and the face often livid, etc.

It is necessary to succor an asphyxiated person with the greatest promptitude, and to continue the remedies as long as there is not a certainty of death. The best and first remedy to employ, and in which the greatest confidence has been and should be placed, is the renewal of the air necessary for respiration. As an instance, let me cite the experience of John Boyle and his son at Yorktown, Penn., who were as near death as men could be, and return to life. The men knew the air was bad in their place (a pitching-breast); but the necessities of life were superior to their discretion, and they continued working until eleven A.M., when the young man concluded he could stand it no longer, and, in making his way from the face to the manway, was overcome, and fell ere he reached it. Mr. Boyle, who was also very weak, took hold of the boy, and, between pulling and lifting him over bowlders, succeeded, very luckily, in reaching the manway; but there he found the air was still too heavy to support life. The exertion and excitement in his endeavor to rescue his boy, and take him down the narrow outlet, together with the deadly gas, proved too much for him, and he, too, fainted away, with the young man in his arms, both becoming tangled fast between the timbers, foot-sills, and slabbing. At five o'clock in the afternoon, a company hand thought something was wrong because of the Boyles remaining in the breast so long, without coming down, and started up the manway after them, and found them near the top, fast, and, as he thought, dead. Aid was summoned, and the miners taken down. The young man revived after reaching the gangway. The father was taken home on a stretcher, and, with much care and labor, was brought back to life. This was a peculiar case. Had there been a larger quantity of black-damp in the air, neither men would have survived; but there seems to have been enough oxygen present to keep life in their bodies, while there was not enough to allow of their keeping their senses.

While it is well enough to know the following methods for restoring asphyxiated persons, and to employ them while waiting for the doctor, yet it is imperative that a physician be summoned immediately.

The following short and clear instructions for the recovery of suffocated persons are those issued by Napoleon, 1813, Tit. iii. Art. xv.

1. Remove the patient to pure air.
2. Undress, and bathe his body with cold water, particularly about the neck.
3. Endeavor to make him swallow, if it be possible, cold water acidulated with vinegar.

4. Clysters should be given, two-thirds of cold water and one-third of vinegar, to be followed with others, of a strong solution of common salt, or senna and epsom salts.

5. Attempts should be made to irritate the pituitary membrane with the feather-end of a quill, which should be gently moved in the nostrils of the patient; or stimulate it with ammonia placed under the nose.

6. Introduce air into the lungs by blowing with the nozzle of a bellows into one of the nostrils, compressing the other with the fingers.

7. If these means do not produce the effects expected, the body of the asphyxiated person remaining warm (as that generally occurs for a long time), it will be necessary to have recourse to blood-letting, the necessity of which is indicated by the redness of the face, the swollen lips, and protruding eyes. The blood may be taken from the jugular vein or foot.

8. As a last resource, an opening should be made in the trachea, and a small pipe introduced, through which the air may be applied by the aid of a small bellows. These remedies should be promptly applied: as death does not appear certain for a long time, they should be only discontinued when it is clearly affirmed.

Although much more might be said about the hygiene of mines, and many rules laid down for the miners and bosses, yet that belongs clearly to another subject; and this would not have been inserted but for the frequency with which such accidents happen, and from the necessity of applying quick remedies. For further information, see "Mine Foreman's Pocket-Book," issued by T. J. Foster, Shenandoah, Penn.

INDEX.

Accidents in mines, 73.
    causes of, 96.
    prevention of, 92.
    decrease of, 93.
Acceleration of gravity, 22.
Action of choke-damp, 11.
    of white-damp, 12.
    of fan and furnace, 112.
Adding airways, 82.
Adjustable fan-shutter, 116.
After-damp, 11.
Air, properties of, 1.
    height of, 2.
    weight of, 3.
    analysis of, 3.
    to find weight of, 19.
    expansion of, 19.
    motion of, 23.
    boxes, 70.
    return, 73.
    stoppings, 74.
    splitting, 76.
    common pressure, in mines, of, 85.

Air for sanitary purposes, 90.
    measurements, 97.
    quantity per horse, 91.
    quantity per lamp, 91.
    quantity per man, 128.
Airways, perimeter of, 42.
    sectional area of, 42.
    different lengths of, 59.
    addition of, 85.
    enlargement of, 92, 97.
Anemometers, 99.
    varieties of, 99.
Aneroid barometer, 105.
Area, 42.
    and quantity, 61.
Arnold, Mr., 98.
Asphyxia, treatment of, 130.
Atkinson, Sir John, 47, 49.
Atmosphere, thickness of, 1.
    pressure of, 2.
    variations of, 27.
Barometer, discovery of, 2.
    correction for, 104.
    use of, 106.

INDEX.

Barometer, sudden fall, 106.
Belgium Commission, 37.
Bends in airways, 41, 79.
Berad, 4.
Black-damp, 9.
Black Hole of Calcutta, 9.
Blood, circulation of, 9.
Blowers, 15, 76.
Blown-out shots, 94.
Boty safety-lamp, 36.
Box, Mr., 129.
Brattice, wooden, 69.
    cloth, 70.
Briam's anemometer, 100.
Buddle, Mr., 30, 76.
Candles in mines, 30.
Carbonic acid in atmosphere, 1.
    exhaled, 8.
    composition of, 9, 90.
    diffusion of, 10.
    properties of, 10.
    sources of, 10.
    fatal nature of, 11.
Carbonic oxide, 12.
    fatal effects of, 13.
Centrifugal ventilators, 110, 115.
Changes of temperature, 28.
Champion ventilator, 111.
Chesterfield and Derbyshire Institute of Mining Engineers, 113.
Choke-damp, 9.
Circular airways, 43.
Clanny safety-lamp, 31.
Coal-dust in dry mines, 95.

Coal-dust explosions, 94.
Co-efficient of friction, 47.
"Colliery Guardian," 94.
Colliery warnings, 108.
Comparative economy of furnace and fan, 117.
Comparison of fans, 111.
    of air and gases, 4.
Composition of air, 3.
    of black-damp, 9.
    of white-damp, 12.
    of sulphuretted hydrogen, 13.
    of fire-damp, 14.
Compressed air in mines, 69.
Contractions in airways, 97.
Correction for anemometer, 100.
    for barometer, 104.
Cost of ventilation, 91, 118.
    by fan, 92, 118.
    by furnace, 92, 118.
Darlington's testimony, 37.
Davy's experiments, 15.
    lamp, 32.
Daubisson, M., 53.
Defective ventilation, 10, 17, 73.
De la Roche, M., 4.
Deputy lamp, 37.
Design of ventilation, 97.
Detection of fire-damp with candle, 30.
Detection of fire-damp with safety-lamp, 39.

INDEX. 137

Diagram of engine, 101.
Dip ventilation, 28.
Discharge through pipes, 54.
    through thin plates, 54.
    of gas, 95.
Distribution of air, 73, 84.
Division of air, 77.
Doors, 74.
Drag in mines, 44.
Drift measurements, 102.
    ventilation, 69.
Dunn, Mr., 129.
Duty of miners, 92.
Economy of furnace and fan, 120.
    of splits, 80.
Effective horse-power, 103.
Efficient ventilation, 96.
Efficiency of fan, 109.
    of furnace, 109.
Eloin safety-lamp, 36.
Enlargement of airways, 92, 97.
Estimating air for mines, 88.
Expansion of gases, 19.
Experiments with anemometer, 99.
    with fan and furnace, 117.
    with fire-damp, 15.
    with safety-lamps, 37, 38.
Explosions, 92.
    indications of (theory), 108.
Face-airing, 71.
Fan, impediments to, 28.

Fan, hand, 69.
    economy of, 92, 118.
    drift, 103, 112.
    Guibal, 115.
    Waddle, 116.
    useful effect of, 116.
    sizes of, 117.
    shutter, 117.
    experiments, 119.
Faraday's estimates, 7.
Fire-damp, 14.
    Davy on, 15.
    discharges of, 16.
    detection of, 39.
Fire-proof brattice-cloth, 70.
Force, estimation of, 44.
Formula for cubic foot of air, 20.
    for total pressure, 50.
    for pressure, 50.
    for motive-column, 26, 51.
    for water-gauge, 52.
    for quantity, 52.
    for units of work, 52.
    for horse-power, 52.
Formulas, 49, 121.
Foster, T. J., 129.
Frankland, Dr., 3.
Free fall, 22.
Friction, 41.
    and area, 43.
    co-efficient of, 47.
    to overcome, 44.
    and pressure, 59.

Friction and power, 55.
    dependent on, 59.
    and velocity, 61.
    and water-gauge, 68, 112.
Furnace, principle of, 24.
    position of, 24.
    power of, 27.
    economy of, 92, 118.
    danger of, 109.
Galileo, 2.
Galloway Royal Society's Journal, 39.
Gases, specific gravity of, 4.
    nitrogen, 5.
    oxygen, 7.
    carbonic acid, 9.
    oxide, 12.
    hydrogen sulphide, 14.
    marsh-gas, 14.
    hydrosulphuric-acid gas, 14.
    proto-carburetted hydrogen, 14.
    light carburetted hydrogen, 14.
    hydride of methyl, 14.
    expansion of, 19.
    chief characteristics, 19.
    Gay-Lussac, 53.
    Goodwin, Mr., 128.
    government warnings, 108.
Gravity, action of, 22.
Guibal, M., experiments of, 99.
    fan experiments, 113.
    fan, 115.

Guibal, M., fan in United States, 117.
    chimney, 116.
Gunpowder, 94.
Gunpowder, charge of, 96.
    use of, 96.
Hall's lamp, 36.
    report, 94.
Hand fans, 69.
Head of air, 25.
Headings, 72.
Heart, action of, 9.
Heat of shafts, 24.
Hedley, Mr., 129.
Horse-power, 52.
    effective, 103.
    nominal, 103.
Hot air, 23.
Howe, R., 113.
Hutchinson, Mr., 128.
Hydrosulphuric-acid gas, 14.
Improvements in safety-lamps, 34.
Illuminating power of, 36.
Improvements in ventilation, 73.
    in fans, 110.
Indicated horse-power, 101.
Indicator-cards, 103.
Inertia, 41.
Influence of air, 5.
Injustice to men, 96.
Inspectors, 92.
Irregular airways, 85.
Large airways, 92.
Laughing-gas, 6.

INDEX. 139

Laws of air in mines, 53.
　of ventilation, 92.
Le Blanc, 10.
Liability of miner, 93.
Limits of velocity, 89.
Lundhill, 73.
Lungs, action of, 8.
　capacity, 130.
Measurements of air by anemometer, 99.
　fan, 101.
　of air, 97.
Mackworth, H., 129.
Magnus, 53.
Managers' ignorance, 92.
Maps of ventilation, 4.
Mariotte's law, 2.
Marsh-gas, 15.
　properties of, 15.
　explosive mixtures, 15.
Mercury, weight of, 15.
Meyer, Dr., 12.
Miners' asthma, 11.
Mines with one orifice, 69.
　with two orifices, 72.
Momentum, 41.
Morison safety-lamp, 38.
Motive-column, 25.
　calculation of, 27.
Movement of air, 23, 43.
Mueseler lamp, 35.
Murphy fan, 111.
Natural ventilation, 23.
Nitrogen, 5.

Nitrogen compounds, 6.
Nixon ventilator, 110.
Nominal horse-power, 103.
North of England Institute, 38, 116.
Oxygen, 7.
　necessity of, 8.
　consumption of, 9.
　starvation, 11.
Orifice of discharge, 54.
Parish safety-lamp, 36.
Peclet, 53.
Pennsylvania mine laws, 129.
Perimeter, 42.
Permanent stoppings, 74.
Phillips, Professor, 129.
Pneumatic paradox, 26.
Poisonous gases, 10, 12, 13.
Position of downcast, 24.
　of fan, 28.
　of upcast, 28.
　of furnace, 24, 28.
Power of winds, 29.
　increase of, 56.
　ventilating, 55.
Pressure of air, 2, 85.
　and water-gauge, 46.
　total, 50.
　per square foot, 50.
　ventilating, 54.
　defined, 55.
　and length, 57.
　and perimeter, 57.
　and area, 58.
　and friction, 59.

140 INDEX.

Pressure and volume, 61.
Prevention of explosions, 92.
Problems, 124.
Properties of air, 40.
Proto-carburetted hydrogen, 14.
Quantity, formula for, 52.
    and area, 61.
    and development, 76.
    for mines, 88.
Rebreathing air, 9, 91.
Regnault, 53.
Regulators, 74, 76.
    to find size of, 83.
Reports of North of England Institute of Mining Engineers, 38.
    of Belgium Commission, 37.
    of Chesterfield and Derbyshire, 113.
    of Mr. Hall, 94.
    of N. Wood, 37.
Resistances to circulation, 23.
Richardson's estimate, 91.
Risca, 73.
Rubbing-surface, 42.
    and friction, 47.
Safety-lamps, discovery of, 31.
    improvements, 34.
    varieties, 34, 93.
    illuminating power of, 36.
    reports on, 37, 38.
    use of, 39.

Scientific ventilation, 88.
Single air currents, 77.
Shaft, ventilation of, 24, 69.
    temperature of, 24.
    position of, 28.
Shot-firing, 96.
Sliding shutters, 76.
Specific gravity, 4.
Spiral fan casing, 117.
Splitting the current, 76.
    advantages of, 79.
    effect of, 82.
South Shields Committee, 37.
Steam-jets, 28, 109.
Steel mill, 31.
Stephenson, Sir George, 32.
    lamp of, 32.
Stoppings, 74.
Struve ventilator, 110.
Stythe, 9.
Sulphuretted hydrogen, 13.
Sulphuric-acid gas, 14.
Sulphuric ether, 98.
Table for specific gravity, 4.
    for weight of air, 21.
    for velocity of winds, 29.
    for water-gauge, 46.
    for square root of water-gauge, 68.
    for air pressure, 107.
    for Guibal fan experiments, 115.
Tarred brattice-cloth, 70.
Temperature, variation of, 2, 27.

INDEX. 141

Temperature and friction, 48.
    and barometer, 106.
Temporary ventilation, 70.
Thick seams, advantage of, 96.
Thomas, J. W., 10, 16, 49.
Throat of fan, 117.
Torricelli, 2.
Total pressure, 50.
Units of work, 52.
Upcast shafts, 24, 28.
Useful effect of air-boxes, 71.
    of engines, 103.
    of fan, 101, 116.
Velocity of air, 20.
    theoretical, 23, 47.
    and head, 47.
    and pressure, 54.
    and friction, 61.
Vena contracta, 54.
Ventilation, 23.
    natural, 24, 69.
    furnace, 24.
    laws of, 53.
    pressure of, 54.
    power, 55, 103.
    shaft, 69.
    drift, 69.
    cost of increasing, 77.
    economy of, 88.

Ventilation, cost of 91.
    of fiery mines, 90.
    arrangements for, 91.
    measurement of, 97.
    mechanical, 110.
    water-fall, 110.
Volume and pressure, 53, 61.
    and length, 59.
    and area, 60.
    and perimeter, 60.
    and rubbing-surface, 60.
Water-gauge, 44.
    use of, 45.
    as a check, 45.
    and pressure, 46, 68.
    formulas for, 52.
    table of, 68.
    and fans, 112.
Weight of mercury, 3.
    of black-damp, 9.
    of fire-damp, 14.
    of air, 19.
    of water, 44.
White-damp, 12.
Williamson's lamp, 36.
Winds an impediment, 28.
    velocity and power of, 29.
Wood, Mr. N., 37.

www.ingramcontent.com/pod-product-compliance
Lightning Source LLC
Chambersburg PA
CBHW031459160426
43195CB00010BB/1028